DR. BARBARA DIABETES COOKBOOK

100 Natural & Delicious Recipes Inspired by Dr. O'Neill to Easily Master Pre-Diabetes and Type 2

Includes a Flavorful 6-Month Meal Plan to Restore Insulin Release

KATHIE GREENE

© Copyright 2024 Kathie Greene

All rights reserved.

The content contained within this book may not be reproduced, duplicated or transmitted without direct written permission from the author or the publisher.

Under no circumstances will any blame or legal responsibility be held against the publisher, or author, for any damages, reparation, or monetary loss due to the information contained within this book, either directly or indirectly.

This book is copyright protected. It is only for personal use. You cannot amend, distribute, sell, use, quote or paraphrase any part, or the content within this book, without the consent of the author or publisher.

By reading this document, the reader agrees that under no circumstances is the author responsible for any losses, direct or indirect, that are incurred as a result of the use of the information contained within this document, including, but not limited to, errors, omissions, or inaccuracies.

TABLE OF CONTENTS

Foreword 1

   About Dr. Barbara O'neill: Her Philosophy And Approach To Holistic Health. 1

Chapter 1 : Introduction 2

Chapter 2 : Diabetes Demystified 4

   Understanding Diabetes: A Simple Overview 4

   Types Of Diabetes: 4

   Understanding Blood Sugar Control: A Simple Guide For Managing Diabetes 5

Chapter 3 : Delightful Mornings: 7

A Fresh Start With Wholesome Breakfasts 7

   Velvety Blueberry Spinach Bliss Smoothie 8

   Creamy Almond Butter & Banana Power Shake 8

   Antioxidant-Rich Green Tea & Kiwi Smoothie 9

   Heart-Healthy Avocado & Mixed Berry Smoothie 9

   Apple Cinnamon Delight Shake 10

   Mango & Coconut Tropical Escape Smoothie 10

   Almond Butter & Mixed Berry Smoothie 11

   Energizing Espresso & Oat Smoothie 11

Wholesome Beginnings With Low-Gi Cereals (For Two) 12

   Crunchy Seed & Nut Morning Muesli 12

   Warm Chia & Golden Hemp Heart Porridge 12

   Buckwheat & Toasted Almond Morning Bowl 13

   Cinnamon Spiced Quinoa Breakfast Cereal 13

   Summer Berries & Amaranth Cereal Bowl 14

   Walnut & Oat Bran Heart-Smart Cereal 14

   Tropical Coconut & Spelt Cereal 15

Pecan & Barley Harvest Bowl 15

Power-Packed Morning Meals (For Two) 16

Savory Tofu & Spinach Scramble 16

Tropical Pineapple & Almond Yogurt Bowl 16

Luxurious Avocado & Chickpea Toast 17

Garden Fresh Veggie & Egg White Frittata 17

Sweet Potato & Black Bean Breakfast Hash 18

Nutty Almond Yogurt & Fresh Fruit Parfait 18

Southwest Tofu & Black Bean Lettuce Wrap 19

Quinoa Power Breakfast Bowl With Soft Boiled Egg 20

No-Fuss Overnight Oats For Busy Mornings (For Two) 21

Cinnamon Apple Pie Overnight Oats 21

Berry Bliss Almond Yogurt Overnight Oats 21

Chapter 4 : Midday Delights: 22

Nutritious And Satisfying Lunches 22

Jackfruit Avocado Quinoa Salad 23

Kale & Roasted Chickpea Delight 24

Sesame Tofu & Crunchy Veggie Bowl 25

Beetroot & Walnut Harmony 26

Broccoli Almond Bliss Salad 27

Nourishing Bowls For Energy-Filled Days (For Two) 28

Lentil & Avocado Power Bowl 28

Mediterranean Hummus & Tofu Salad 29

Vegan & Savory Black Bean Wrap 30

Tropical Tofu & Fruit Salad 31

Herbed Egg & Veggie Fiesta 32

Flavorful Wraps & Wholesome Sandwiches: Easy And Enticing Lunchtime Solutions (For Two) 33

Homemade Whole Grain Tortilla Recipe 33

Avocado & Quinoa Salad Wrap 34

Zesty Black Bean And Corn Wrap 34

Chickpea Caesar Wrap 35

Roasted Pepper And Hummus Wrap 36

Grilled Eggplant And Fresh Pesto Panini 37

Vegan Lentil Salad Sandwich On Whole Grain Bread 38

Homemade Veggie Burger Sandwich 39

Chickpea Salad Sandwich 39

Soul-Soothing Soup Symphony (For Two) 40

Homemade Vegetable Broth 40

Roasted Tomato Basil Soup 40

Creamy Carrot And Ginger Soup 41

Savory Chickpea And Vegetable Soup 42

Lentil And Spinach Soup 43

Mushroom And Barley Soup 43

Sweet Potato And Coconut Soup 44

Simple Jackfruit And Wild Rice Soup 45

Vegan Corn And Potato Chowder 46

Green Gourmet: Plant-Based Dishes For Every Palate (For Two) 47

Stuffed Bell Peppers With Quinoa And Black Beans 47

Vegan Shepherd's Pie With Lentils 48

Zucchini Noodle Pad Thai 49

Cauliflower Steak With Chimichurri Sauce 50

Mushroom Stroganoff With Cashew Cream 51

Vegan Jambalaya With Jackfruit 51

Spaghetti Squash With Tomato Basil Sauce 52

Ratatouille With Herbed Polenta 52

Chapter 5 : Nourishing Dinners: Wholesome Evening Meals 53

Grilled Tofu With Lemon-Dill Sauce 54

Herb-Roasted Cauliflower Steaks 54

Seared Tofu Steaks With Avocado-Wasabi Sauce 55

Vegan Lentil Meatballs In Tomato Basil Sauce 56

Quinoa And Black Bean Stuffed Bell Peppers 57

Baked Tofu With Mango Salsa 58

Vegetable Stir-Fry With Black Beans, Broccoli And Bell Peppers 59

Grilled Portobello Mushrooms With Lentil Mint Pesto 59

Healthy Sides: Perfect Pairings For A Balanced Meal (For Two) 60

Quinoa Salad With Cucumber And Avocado 60

    Steamed Asparagus With Toasted Almonds 60

    Mashed Sweet Potatoes With Cinnamon 61

    Roasted Brussels Sprouts With Balsamic Reduction 61

    Carrot And Zucchini Ribbons With Lemon Vinaigrette 62

    Garlic Cauliflower "Rice" 62

    Sautéed Spinach With Pine Nuts And Raisins 63

    Grilled Corn On The Cob With Chili Lime Dressing 63

Chapter 6 : Healthy Indulgences: 64

Satisfying Snacks For Every Occasion 64

    Spiced Chickpea Nuts 65

    Zucchini Chips 65

    Cucumber Cups With Herbed Almond Yogurt 66

    Avocado Hummus 66

    Baked Kale Chips 67

    Roasted Pumpkin Seeds 67

    Stuffed Bell Peppers With Quinoa And Vegetables 68

    Herb And Garlic Mushroom Caps 68

Sweet Sensations: Naturally Delightful (For Two) 69

    Cinnamon Flaxseed Pudding 69

    Fresh Berry Salad 69

    Apple Cinnamon Chips 70

    Chia And Berry Parfait 70

    Nutty Stuffed Pears 71

    Sweet Cinnamon Almond Mix 71

    Avocado Vanilla Mousse 72

    Carrot Cake Balls 72

Grab-And-Go Treats: Quick And Convenient (For Two) 73

    Almond Butter Energy Balls 73

    Savory Roasted Chickpeas 73

    Cinnamon Nut Snack Mix 74

    Flaxseed And Blueberry Parfait 74

    Pumpkin Seed And Sunflower Seed Mix 75

    Avocado And Tomato Cucumber Cups 75

    Spiced Pear Slices 76

Sweet Potato And Almond Butter Slices (Baked) 76

Chapter 7 : Exclusive Bonuses 77

1. Hydration And Drinks: Secrets By Barbara O'neill 77

2. "Ayurveda For Women": The Ultimate Wellness Guide 77

3. Healing With Nature: Barbara O'neill's Natural Remedies 77

4. Living Well With Diabetes: Exercise And Stress Reduction 77

5. Cure For High Blood Pressure: Inspired By Barbara O'neill 77

Chapter 8 : 6-Month Meal Plan 78

Week 1 78

Week 2 79

Week 3 80

Week 4 81

Week 5 82

Week 6 83

Week 7 84

Week 8 85

Week 9 87

Week 10 88

Week 11 89

Week 12 90

Week 13 91

Week 14 92

Week 15 93

Week 16 94

Week 17 95

Week 18 96

Week 19 98

Week 20 99

- Week 21 100
- Week 22 101
- Week 23 102
- Week 24 103
- Week 25 104
- Week 26 105
- Week 27 107
- Week 28 108

Chapter 9 110

Cooking Conversions 110

Conclusion 112

Analytic Index 114

FOREWORD

About Dr. Barbara O'Neill: Her philosophy and approach to holistic health.

Step into the light of holistic health with Dr. Barbara O'Neill, a pioneer in integrating natural health solutions with traditional medical understanding. With over three decades of expertise in naturopathy and public health, Dr. O'Neill isn't just a practitioner; she's a guide leading thousands to a more vibrant way of living through the power of informed dietary choices and comprehensive lifestyle changes.

Raised in a health-conscious family, Dr. O'Neill early on embraced the potent healing powers of nature. As a respected naturopathic doctor and public health educator, her approach goes beyond the typical treatment of symptoms. She delves deep, advocating a life where managing diabetes means enhancing overall well-being—not through prescriptions alone but through nourishing meals, purposeful movement, and mental resilience.

Dr. O'Neill's philosophy centers on the body's intrinsic ability to heal itself, given the right conditions. Her practical application of this philosophy is evident in her passionate teaching and the transformative impact she has on those she reaches. Through her work, Dr. O'Neill doesn't just change diets; she changes lives, empowering individuals with the knowledge and tools to achieve better health and a deeper joy in life.

CHAPTER 1
INTRODUCTION

In this life, where every meal can feel like a battlefield against diabetes, a guide like this cookbook can be your map and compass, helping you navigate through the complex terrain of blood sugar management and wellness. Here lies not just a collection of recipes but a manifesto for change, empowering you to reclaim control over your diet and, in turn, over your diabetes.

Understanding the Purpose

This cookbook's mission is profound yet simple: to empower you. Managing diabetes effectively requires more than just monitoring your blood sugar; it involves nurturing your entire being through informed dietary choices and structured meal plans. Here, you will learn that food is not merely sustenance but medicine, capable of healing and enhancing your wellbeing. The recipes and meal plans within these pages are designed not only to keep your blood sugar in check but also to enrich your overall health, offering a pathway to a fuller, more vibrant life.

Getting Started

Dive into this culinary journey by first assessing where you stand. Look at your current eating habits—what you eat, when you eat, and how you prepare your food. This cookbook provides you with the tools to understand the nutritional impact of your food choices and guides you on how to equip your kitchen with the essentials. Ready your space and your mind for this transformative experience. Embrace the changes with openness and readiness, both mentally and physically, as you step into a new way of eating and living.

Navigating the Cookbook

Let's take a brief tour through the layout of this cookbook. Organized thoughtfully into sections—breakfasts, lunches, dinners, snacks, and desserts—each part caters to the unique nutritional needs of individuals managing diabetes. The structure is intuitive, designed to help you easily locate recipes and information. Use the index and thematic tags to navigate swiftly to the foods that interest you, making meal preparation an enjoyable and stress-free experience.

Personalizing Meal Plans

Flexibility is key in managing diabetes, and this cookbook celebrates that. Here, you are encouraged to tailor recipes to suit your taste preferences, dietary restrictions, and nutritional needs. Feel free to swap ingredients, try new flavor combinations, and adjust portions to maintain the necessary balance for effective diabetes management. This section of the cookbook encourages creativity and personalization, ensuring that your meal plan is as unique as you are.

Practical Usage of Recipes

Each recipe is crafted not only with taste in mind but with a clear focus on health benefits—whether it's a low glycemic index, high fiber content, or richness in omega-3 fatty acids. Understand why these elements are crucial and how they contribute to managing your diabetes effectively. This cookbook encourages you to see these recipes as flexible guides, adaptable based on the seasonal availability of produce or your current health goals.

Tracking Progress and Adjustments

To truly gauge the effectiveness of your dietary changes, it's vital to monitor your blood sugar levels and other health indicators. This cookbook suggests keeping a detailed food diary or using digital tools to track your progress. Regular checkpoints are recommended to assess how well the diet is working and to make necessary adjustments, ensuring that your meal plan evolves with your health needs.

Building a Support System

Changing lifelong eating habits is no small feat and having a robust support system is crucial. Whether it's involving family and friends, connecting with online communities, or joining local groups with similar health goals, this cookbook emphasizes the importance of support in your journey. You're not alone, and building relationships with those who understand and support your goals can be incredibly empowering.

Continuous Learning and Adaptation

Managing diabetes is an ongoing journey of discovery and adaptation. This cookbook encourages you to stay curious and informed about the latest research in diabetes care and nutrition. Keep an open mind, ready to adapt your approach as you learn more about how your body responds to different foods and treatments.

Encouragement and Inspiration

Finally, let this cookbook serve as a source of encouragement and inspiration. Embrace these changes not as burdens but as exciting opportunities to enhance your quality of life through thoughtful, informed dietary choices. This cookbook isn't just about recipes; it's a tool for transformation and empowerment. Let it inspire you to take control of your diabetes with hope and confidence, reminding you that every meal is a step towards a healthier, happier you.

CHAPTER 2
DIABETES DEMYSTIFIED

Understanding Diabetes: A Simple Overview

Diabetes is a chronic condition that fundamentally affects how your body converts food into energy. Most of the food you eat is broken down into sugar (also called glucose) and released into your bloodstream. When blood sugar levels rise, it signals your pancreas to release insulin.

What is Insulin?

Insulin is a hormone produced by the pancreas that acts like a key to allow blood sugar into your body's cells for use as energy. In diabetes, your body either doesn't make enough insulin or can't effectively use the insulin it produces. This results in too much blood sugar staying in your bloodstream, which over time can lead to serious health problems such as heart disease, vision loss, and kidney disease.

Types of Diabetes:

1. Type 1 Diabetes

- **What It Is:** An autoimmune disease where the immune system mistakenly attacks and destroys the insulin-producing cells in the pancreas.
- **How It Arises:** Not related to lifestyle; generally appears in childhood or adolescence.
- **Management:** Requires regular insulin injections or an insulin pump to maintain blood glucose levels.

2. Type 2 Diabetes

- **What It Is:** The most common form of diabetes, where the body becomes resistant to insulin or doesn't produce enough insulin.
- **How It Arises:** Often linked to lifestyle factors such as obesity, physical inactivity, and poor dietary choices, though genetics can also play a role.
- **Management:** Often managed or even reversed through lifestyle changes such as improved diet, increased physical activity, and weight loss, along with medication if necessary.

3. Gestational Diabetes

- **What It Is:** Diabetes that develops during pregnancy and typically resolves after giving birth.
- **How It Arises:** Caused by hormonal changes during pregnancy that make the body more resistant to insulin.
- **Management:** Managed through diet and careful monitoring of blood glucose levels to prevent complications during pregnancy. It is important to monitor for Type 2 diabetes development later in life.

Each type of diabetes has its unique challenges and requires different management strategies. Understanding these differences is crucial in taking informed steps towards effective diabetes management and maintaining overall health. This guide aims to provide clear, empathetic information to help you or your loved ones manage diabetes with knowledge and confidence.

4. Pre-Diabetes

- **What It Is:** A condition where blood sugar levels are higher than normal but not yet high enough to be diagnosed as Type 2 diabetes.
- **How It Arises:** Often precedes Type 2 diabetes, linked to similar lifestyle factors such as excess weight and inactivity.
- **Management:** Lifestyle changes such as diet improvement and increased physical activity are crucial to prevent progression to Type 2 diabetes.

Understanding Blood Sugar Control: A Simple Guide for Managing Diabetes

Managing diabetes effectively revolves around keeping your blood sugar levels within a healthy range. This is what doctors call "glycemic control," and it's key to avoiding the complications often associated with diabetes, such as heart problems and nerve damage. Let's break down what this really means in everyday terms.

What Does Glycemic Control Mean?

Glycemic control is about balancing your blood sugar (glucose) levels. If you have diabetes, your doctor will likely recommend keeping your blood sugar levels within these general targets:

- **Before meals:** 80-130 mg/dL
- **Two hours after meals:** Below 180 mg/dL

Staying within these ranges can help you avoid long-term health issues and feel your best each day.

Understanding Glucose and Insulin

- **Glucose as Energy:** Whenever you eat, your body breaks down foods into glucose, which is a main source of energy. This glucose goes into your bloodstream and raises your blood sugar levels.
- **The Role of Insulin:** Insulin is like a key that helps open your body's cells so they can take in glucose. If you have diabetes, your body might not make enough insulin or use it well, leading to higher blood sugar levels.

The Glycemic Index (GI)

The glycemic index helps us understand how different foods affect blood sugar levels:

- **Low GI Foods (55 or less)** like apples and lentils, increase your blood sugar slowly and are better for keeping it stable.

- **Medium GI Foods (56-69)** like brown rice, have a moderate effect on blood sugar.
- **High GI Foods (70 or more)** like white bread, raise your blood sugar quickly and should be eaten less often.

Monitoring Your Blood Sugar

Keeping track of your blood sugar levels helps you manage diabetes effectively:

- **Daily Checks:** Regular testing lets you see if you're staying within your target range and helps you understand how different foods and activities affect your blood sugar.
- **Tools to Help:** You can use a blood glucose meter for spot checks throughout the day or wear a continuous glucose monitor (CGM) to get real-time readings.

Dealing with High and Low Blood Sugar

High Blood Sugar (Hyperglycemia):

- **Symptoms:** Feeling very thirsty, tired, or needing to urinate more often.
- **What to Do:** You might need to adjust what you eat or your medication. Always consult with your healthcare provider for guidance.

Low Blood Sugar (Hypoglycemia):

- **Symptoms:** Feeling shaky, dizzy, or very hungry.
- **What to Do:** Eating something that has about 15-20 grams of simple carbs (like fruit juice or glucose tablets) can help bring your levels back to normal quickly.

We are committed to creating high-quality products and aim for complete customer satisfaction. Your support in our growth would mean a great deal to us. After finishing the book, we would greatly appreciate your feedback on Amazon, as your thoughts are extremely valuable to us. Thank you wholeheartedly!

CHAPTER 3
DELIGHTFUL MORNINGS:

A FRESH START WITH WHOLESOME BREAKFASTS

SHAKES AND SMOOTHIES TO ENERGIZE YOUR DAY (FOR TWO)

VELVETY BLUEBERRY SPINACH BLISS SMOOTHIE

Ingredients

- 1 cup fresh spinach
- 1/2 cup frozen blueberries
- 1 banana
- 1 tbsp. chia seeds
- 1 cup unsweetened almond milk
- 1 tsp stevia (optional, adjust to taste)

Nutritional Information (per serving)

- Calories: 120
- Carbohydrates: 22g
- Proteins: 2g
- Fats: 1g

Estimated Glycemic Index: Low

Instructions:

1. Combine all ingredients in a blender.
2. Blend until smooth and creamy.
3. Serve immediately, offering a refreshing and nutritious beverage perfect for two.

CREAMY ALMOND BUTTER & BANANA POWER SHAKE

Ingredients

- 1 ripe banana
- 2 tbsp. almond butter
- 1 cup homemade almond milk thickened with chia seeds (see instructions below)
- 1/2 cup almond milk
- 1 tsp stevia (optional, adjust to taste)

Nutritional Information (per serving)

- Calories: 300
- Carbohydrates: 33g
- Proteins: 8g
- Fats: 16g

Estimated Glycemic Index: Medium

Instructions:

1. Prepare Thickened Almond Milk: To make one cup of thickened almond milk, blend 1 cup of homemade almond milk with 2 tablespoons of chia seeds. Let it sit for about 30 minutes until the mixture reaches a thick, yogurt-like consistency.
2. Blend Ingredients: Place the ripe banana, almond butter, thickened almond milk, regular almond milk, and optional stevia in a blender.
3. Blend until Creamy: Blend all the ingredients until smooth and creamy.
4. Serve: Pour the shake into two glasses and enjoy as a satisfying, protein-rich beverage.

ANTIOXIDANT-RICH GREEN TEA & KIWI SMOOTHIE

Ingredients
- 1 cup brewed green tea, cooled
- 2 ripe kiwis, peeled and sliced
- 1/2 apple, cored and chopped
- 1 tsp stevia (optional, adjust to taste)

Nutritional Information (per serving)
- Calories: 100
- Carbohydrates: 23g
- Proteins: 1g
- Fats: 0.5g

Estimated Glycemic Index: Low

Instructions:
1. Combine the green tea, kiwi slices, chopped apple, and stevia in a blender.
2. Blend until smooth.
3. Serve chilled, dividing the smoothie between two glasses for a powerful antioxidant boost.

HEART-HEALTHY AVOCADO & MIXED BERRY SMOOTHIE

Ingredients
- 1/2 ripe avocado
- 1/2 cup mixed berries (strawberries, blueberries, raspberries), frozen
- 1 cup spinach leaves
- 1 cup unsweetened almond milk
- 1 tsp stevia (optional, adjust to taste)

Nutritional Information (per serving)
- Calories: 220
- Carbohydrates: 15g
- Proteins: 3g
- Fats: 15g

Estimated Glycemic Index: Low

Instructions:
1. Add all ingredients to a blender.
2. Blend until smooth and creamy.
3. Serve immediately, providing a heart-healthy smoothie for two.

APPLE CINNAMON DELIGHT SHAKE

Ingredients

- 1 apple, cored and sliced
- 1 cup homemade vanilla-flavored almond milk thickened with chia seeds (see instructions below)
- 1/2 tsp ground cinnamon
- 1/2 cup ice
- 1 tsp stevia (optional, adjust to taste)

Nutritional Information (per serving)

- Calories: 145
- Carbohydrates: 25g
- Proteins: 4g
- Fats: 2g

Estimated Glycemic Index: Low

Instructions:

1. Prepare Thickened Vanilla Almond Milk: To make one cup, blend 1 cup of homemade almond milk with 1 tbsp. of chia seeds and 1/2 tsp of natural vanilla extract. Let it sit for about 30 minutes or until the mixture reaches a thick, yogurt-like consistency.
2. Blend Ingredients: Place the apple slices, thickened vanilla almond milk, ground cinnamon, ice, and optional stevia in a blender.
3. Blend until Smooth and Creamy: Blend all the ingredients until smooth.
4. Serve: Pour the shake into glasses and serve immediately. Enjoy this crisp and comforting beverage that's perfect for a refreshing treat.

MANGO & COCONUT TROPICAL ESCAPE SMOOTHIE

Ingredients

- 1 mango, peeled and diced
- 1/2 cup coconut milk
- 1/2 banana
- 1/2 cup crushed ice
- 1 tsp stevia (optional, adjust to taste)

Nutritional Information (per serving)

- Calories: 200
- Carbohydrates: 30g
- Proteins: 3g
- Fats: 8g

Estimated Glycemic Index: Medium

Instructions:

1. Place the diced mango, coconut milk, banana, and ice in a blender.
2. Add stevia to taste.
3. Blend until smooth. Pour into glasses and transport yourself to a tropical paradise with each sip.

ALMOND BUTTER & MIXED BERRY SMOOTHIE

Ingredients

- 1/2 cup mixed berries (blueberries, raspberries, strawberries – fresh or frozen)
- 2 tbsp almond butter
- 1/2 cup thickened vanilla almond milk
- 1/2 banana
- 3 tbsp hemp seeds
- 1 tsp stevia (optional, adjust to taste)

Nutritional Information (per serving)

- Calories: 280
- Carbohydrates: 21g
- Proteins: 15g
- Fats: 17g

Estimated Glycemic Index: Low

Instructions:

1. Prepare Thickened Vanilla Almond Milk: Mix 1 cup of almond milk with 1 tbsp of chia seeds and 1/2 tsp of vanilla extract. Let it sit for about 30 minutes until it thickens.
2. Blend Ingredients: In a blender, combine the mixed berries, almond or sunflower seed butter, thickened vanilla almond milk, banana, hemp seeds, and optional stevia.
3. Combine strawberries, peanut butter, Greek yogurt, banana, and stevia in a blender.
4. Blend all ingredients until smooth. The thickness of the almond milk will help emulate a creamy, rich texture.
5. Serve: Pour the smoothie into glasses and enjoy as a filling breakfast or a nutritious snack.

ENERGIZING ESPRESSO & OAT SMOOTHIE

Ingredients

- 1/2 cup rolled oats
- 1 shot espresso, cooled
- 1 banana
- 1 cup almond milk
- 1 tsp stevia (optional, adjust to taste)

Nutritional Information (per serving)

- Calories: 240
- Carbohydrates: 35g
- Proteins: 8g
- Fats: 5g

Estimated Glycemic Index: Low

Instructions:

1. Soak rolled oats in almond milk for 10 minutes to soften.
2. Add softened oats, cooled espresso, banana, and stevia to a blender.
3. Blend until smooth. This smoothie is perfect for those mornings when you need an extra boost to start your day energetically.

WHOLESOME BEGINNINGS WITH LOW-GI CEREALS (FOR TWO)

CRUNCHY SEED & NUT MORNING MUESLI

Ingredients
- 1 cup whole oats
- 1/4 cup mixed nuts (almonds, walnuts, cashews), chopped
- 2 tbsp sunflower seeds
- 2 tbsp pumpkin seeds
- 1/2 cup fresh berries
- Unsweetened almond milk (for serving)

Nutritional Information (per serving)
- Calories: 300
- Carbohydrates: 45g
- Proteins: 10g
- Fats: 15g

Estimated Glycemic Index: Low

Instructions:
1. In a large mixing bowl, combine whole oats, chopped mixed nuts, sunflower seeds, pumpkin seeds, and chia seeds. Mix thoroughly to ensure even distribution.
2. To serve, scoop the desired amount of muesli into a bowl.
3. Pour unsweetened almond milk over the muesli to your liking.
4. Top with fresh berries of your choice for added freshness and a burst of flavor.
5. Stir the mixture gently to combine the flavors before enjoying a nutritious start to your day.

WARM CHIA & GOLDEN HEMP HEART PORRIDGE

Ingredients
- 1/2 cup rolled oats
- 2 tbsp chia seeds
- 2 tbsp hemp hearts
- 1 tbsp ground flaxseed
- Almond milk (enough to cover the oats in the saucepan)
- 1 tbsp stevia (adjust to taste)
- Optiona: cinnamom (to taste)

Nutritional Information (per serving)
- Calories: 250
- Carbohydrates: 30g
- Proteins: 15g
- Fats: 14g

Estimated Glycemic Index: Low

Instructions:
1. Combine rolled oats, chia seeds, hemp hearts, flaxseed, and almond milk in a saucepan.
2. Cook over medium heat, stirring frequently until the porridge thickens.
3. Stir in cinnamon (if using) and stevia for sweetness.
4. Serve warm for a hearty, nutritious breakfast.

BUCKWHEAT & TOASTED ALMOND MORNING BOWL

Ingredients
- 1/2 cup buckwheat groats
- 1/4 cup almonds, toasted and chopped
- 1 tbsp stevia (adjust to taste)
- 1 cup low-fat milk or almond milk

Nutritional Information (per serving)
- Calories: 200
- Carbohydrates: 33g
- Proteins: 6g
- Fats: 8g

Estimated Glycemic Index: Low

Instructions:
1. Rinse buckwheat groats under cold water.
2. Place the rinsed buckwheat in a small saucepan and add low-fat milk or almond milk.
3. Bring the mixture to a boil over medium heat, then reduce the heat to low and simmer uncovered, stirring occasionally, until the buckwheat is tender and most of the liquid has been absorbed, about 10 minutes.
4. Stir in toasted chopped almonds, stevia, and vanilla extract.
5. Serve warm, a perfect blend of nutty flavors and whole grains.

CINNAMON SPICED QUINOA BREAKFAST CEREAL

Ingredients
- 1/2 cup quinoa, rinsed
- 1 cup water
- 1 tsp cinnamon
- 1/4 cup raisins
- 1 tbsp stevia (adjust to taste)
- 1/2 apple, diced

Nutritional Information (per serving)
- Calories: 220
- Carbohydrates: 39g
- Proteins: 8g
- Fats: 3.5g

Estimated Glycemic Index: Low

Instructions:
1. Combine quinoa, water, and cinnamon in a saucepan. Bring to a boil.
2. Reduce heat to low, cover, and simmer until quinoa is tender and water is absorbed, about 15 minutes.
3. Stir in raisins, stevia, and diced apple.
4. Serve hot for a delicious and spiced morning meal.

SUMMER BERRIES & AMARANTH CEREAL BOWL

Ingredients
- 1/2 cup amaranth
- 1 cup water
- 1/2 cup mixed summer berries (blueberries, strawberries, raspberries)
- 1 tbsp chia seeds

Nutritional Information (per serving)
- Calories: 250
- Carbohydrates: 46g
- Proteins: 7g
- Fats: 5g

Estimated Glycemic Index: Low

Instructions:
1. Rinse amaranth under cold water until water runs clear.
2. In a pot, bring water to a boil, add amaranth, reduce heat to low, cover, and simmer for 20 minutes.
3. Once cooked, stir in almond milk, stevia, and chia seeds.
4. Top with mixed summer berries and serve warm. This bowl combines the hearty texture of amaranth with the freshness of berries for a perfect start.

WALNUT & OAT BRAN HEART-SMART CEREAL

Ingredients
- 1/2 cup oat bran
- 1/4 cup walnuts, chopped
- 1 tbsp flaxseed meal
- 1 cup almond milk
- 1 tsp stevia (adjust to taste)

Nutritional Information (per serving)
- Calories: 150
- Carbohydrates: 18g
- Proteins: 6g
- Fats: 20g

Estimated Glycemic Index: Low

Instructions:
1. In a saucepan, combine oat bran with almond milk and bring to a simmer over medium heat.
2. Cook while stirring until thickened, about 5 minutes.
3. Remove from heat and stir in chopped walnuts, flaxseed meal, and stevia.
4. Serve warm for a heart-healthy, filling breakfast.

TROPICAL COCONUT & SPELT CEREAL

Ingredients
- 1/2 cup spelt flakes
- 1/4 cup shredded coconut
- 1 cup unsweetened coconut milk
- 1/2 banana, sliced
- 1 tsp stevia (adjust to taste)

Nutritional Information (per serving)
- Calories: 180
- Carbohydrates: 27g
- Proteins: 4g
- Fats: 12g

Estimated Glycemic Index: Medium

Instructions:
1. In a small pot, heat coconut milk and add spelt flakes.
2. Cook over medium heat until the spelt is tender, about 10 minutes.
3. Stir in shredded coconut and stevia for sweetness.
4. Top with banana slices before serving. Enjoy a tropical-inspired, nourishing breakfast.

PECAN & BARLEY HARVEST BOWL

Ingredients
- 1/2 cup hulled barley
- 1 cup water
- 1/4 cup pecans, toasted and chopped
- 1 apple, diced
- 1 tsp cinnamon
- 1 tbsp stevia (adjust to taste)

Nutritional Information (per serving)
- Calories: 210
- Carbohydrates: 44g
- Proteins: 6g
- Fats: 9g

Estimated Glycemic Index: Low

Instructions:
1. Rinse barley under cold water, then combine with water in a pot.
2. Bring to a boil, reduce heat to low, cover, and simmer until barley is tender and water is absorbed, about 30-40 minutes.
3. Stir in toasted pecans, diced apple, cinnamon, and stevia.
4. Serve warm for a hearty and satisfying start to your day.

POWER-PACKED MORNING MEALS (FOR TWO)

SAVORY TOFU & SPINACH SCRAMBLE

Ingredients
- 1/2 block firm tofu, crumbled
- 1 cup fresh spinach, chopped
- 1/4 cup chopped onions
- 1 clove garlic, minced
- 1 tbsp olive oil
- 1/2 tsp turmeric (for color)
- Salt and pepper to taste

Nutritional Information (per serving)
- Calories: 180
- Carbohydrates: 4g
- Proteins: 20g
- Fats: 9g

Estimated Glycemic Index: Low

Instructions:
1. Heat olive oil in a skillet over medium heat. Add onions and garlic, sauté until translucent.
2. Add crumbled tofu and turmeric, stir to combine. Cook for about 5 minutes until the tofu is golden.
3. Stir in chopped spinach and cook until wilted.
4. Season with salt and pepper. Serve warm for a hearty, protein-rich breakfast.

TROPICAL PINEAPPLE & ALMOND YOGURT BOWL

Ingredients
- 1 cup almond yogurt (unsweetened)
- 1/2 cup chopped fresh pineapple
- 1/4 cup sliced almonds
- 1 tbsp shredded coconut (unsweetened)

Nutritional Information (per serving)
- Calories: 220
- Carbohydrates: 22g
- Proteins: 5g
- Fats: 14g

Estimated Glycemic Index: Medium

Instructions:
1. In a bowl, combine almond yogurt with chopped pineapple.
2. Top with sliced almonds and shredded coconut for a tropical flavor.
3. Mix gently and enjoy a refreshing and protein-packed breakfast bowl.

LUXURIOUS AVOCADO & CHICKPEA TOAST

Ingredients
- 2 slices whole grain bread, toasted
- 1 cup cooked chickpeas
- 1 ripe avocado, mashed
- 1 tbsp capers
- 1 tsp lemon juice
- Salt and pepper to taste

Nutritional Information (per serving)
- Calories: 300
- Carbohydrates: 38g
- Proteins: 12g
- Fats: 15g

Estimated Glycemic Index: Low

Instructions:
1. Prep the Chickpeas: Mash the cooked chickpeas slightly with a fork to add texture.
2. Prepare the Avocado Spread: In a small bowl, mash the ripe avocado until creamy. Mix in lemon juice, and season with salt and pepper to taste.
3. Assemble the Toast: Spread the mashed avocado evenly over the toasted whole grain bread slices. Top with mashed chickpeas and sprinkle with capers.
4. Serve: Serve immediately for a luxurious and satisfying breakfast.

GARDEN FRESH VEGGIE & EGG WHITE FRITTATA

Ingredients
- 4 egg whites
- 1/2 cup diced bell peppers
- 1/4 cup diced tomatoes
- 1/4 cup chopped zucchini
- 1 tbsp chopped fresh basil
- 1 tsp olive oil
- Salt and pepper to taste

Nutritional Information (per serving)
- Calories: 140
- Carbohydrates: 8g
- Proteins: 16g
- Fats: 5g

Estimated Glycemic Index: Low

Instructions:
1. Preheat oven to 375°F (190°C).
2. In a skillet, heat olive oil over medium heat. Sauté bell peppers, tomatoes, and zucchini until soft.
3. In a bowl, whisk egg whites with salt, pepper, and basil.
4. Pour the egg mixture over the veggies in the skillet. Cook for 2-3 minutes until the edges start to set.
5. Transfer skillet to oven and bake for 8-10 minutes until the frittata is set.
6. Serve warm, garnished with additional basil if desired. This vegetable-packed frittata is light, flavorful, and nutritious.

SWEET POTATO & BLACK BEAN BREAKFAST HASH

Ingredients
- 1 medium sweet potato, diced
- 1 cup cooked black beans
- 1/2 red bell pepper, diced
- 1/2 onion, diced
- 2 tbsp olive oil
- Salt and pepper to taste

Nutritional Information (per serving)
- Calories: 270
- Carbohydrates: 35g
- Proteins: 8g
- Fats: 12g

Estimated Glycemic Index: Low

Instructions:
1. Heat olive oil in a large skillet over medium heat.
2. Add diced sweet potato and cook until starting to soften, about 5 minutes.
3. Add diced red bell pepper, onion, and cooked black beans to the skillet. Cook until the vegetables are tender and the beans are heated through, about 10 minutes.
4. Season with salt and pepper to taste.
5. Serve hot for a hearty, nutritious start to your day with this delicious black bean and sweet potato hash.

NUTTY ALMOND YOGURT & FRESH FRUIT PARFAIT

Ingredients
- 1 cup almond yogurt (unsweetened)
- 1/4 cup pecans, toasted
- 1/2 cup fresh berries (strawberries, blueberries, raspberries)
- 1/2 tsp cinnamon
- 1 tbsp chia seeds

Nutritional Information (per serving)
- Calories: 180
- Carbohydrates: 16g
- Proteins: 6g
- Fats: 12g

Estimated Glycemic Index: Low

Instructions:
1. Toast the pecans: Lightly toast the pecans in a dry skillet over medium heat until fragrant.
2. Prepare the parfait layers: In a serving glass, layer half of the almond yogurt at the bottom.
3. Add fruits and nuts: Add a layer of fresh berries, then sprinkle some toasted pecans and a pinch of cinnamon over them.
4. Add chia seeds: Sprinkle a tablespoon of chia seeds for an extra nutritional boost.
5. Repeat the layers: Layer the remaining yogurt, followed by another layer of berries, pecans, and a final sprinkle of cinnamon.
6. Chill if desired: Though ready to eat immediately, this parfait can also be refrigerated for an hour to allow the flavors to meld together.

SOUTHWEST TOFU & BLACK BEAN LETTUCE WRAP

Ingredients

- 2 large lettuce leaves or collard greens (whole leaves, blanched if preferred for pliability)
- 1/2 cup scrambled tofu
- 1/4 cup black beans
- 1/4 cup shredded lettuce or additional greens for filling

For the salsa:

- 1 small tomato, finely diced
- 1/4 small red onion, finely diced
- 1/2 jalapeño pepper, finely diced (optional, adjust to taste)
- 1 tbsp fresh cilantro, chopped Juice of 1/2 lime
- Salt and pepper to taste

Nutritional Information (per serving)

- Calories: 200
- Carbohydrates: 15g
- Proteins: 12g
- Fats: 10g

Estimated Glycemic Index: Low

Instructions:

1. Prepare the Salsa: In a small bowl, combine the diced tomato, red onion, jalapeño pepper (if using), chopped cilantro, lime juice, salt, and pepper. Mix well.
2. Prepare the Lettuce or Collard Greens: If using collard greens, blanch them briefly in boiling water to make them more pliable. Rinse under cold water and pat dry.
3. Add Filling: Lay the lettuce or collard green leaves flat on a plate. Spoon the scrambled tofu and black beans onto the leaves. Top with shredded lettuce or additional greens.
4. Add Salsa: Spoon the freshly made salsa over the tofu and beans.
5. Roll the Leaves: Carefully roll the leaves to enclose the filling, tucking in the edges if possible to secure the wrap.
6. Serve: Enjoy immediately for a fresh, wholesome, and satisfying meal..

QUINOA POWER BREAKFAST BOWL WITH SOFT BOILED EGG

Ingredients

- 1/2 cup cooked quinoa
- 1 soft boiled egg
- 1/4 cup diced avocado
- 1 tbsp pumpkin seeds
- 1/4 cup chopped spinach
- Salt and pepper to taste

Nutritional Information (per serving)

- Calories: 250
- Carbohydrates: 20g
- Proteins: 15g
- Fats: 15g

Estimated Glycemic Index: Low

Instructions:

1. Prepare a soft boiled egg to your liking.
2. In a bowl, combine cooked quinoa, diced avocado, pumpkin seeds, and chopped spinach.
3. Peel the soft boiled egg and place on top of the quinoa mixture.
4. Season with salt and pepper.
5. Enjoy a nutrient-rich, energizing breakfast bowl.

NO-FUSS OVERNIGHT OATS FOR BUSY MORNINGS (FOR TWO)

CINNAMON APPLE PIE OVERNIGHT OATS

Ingredients
- 1/2 cup rolled oats
- 1 small apple, diced
- 1/2 tsp cinnamon
- 1 tbsp chia seeds
- 1 cup unsweetened almond milk
- 1 tsp stevia (adjust to taste)

Nutritional Information (per serving)
- Calories: 310
- Carbohydrates: 55g
- Proteins: 8g
- Fats: 7g

Estimated Glycemic Index: Low

Instructions:
1. In a mason jar or airtight container, combine rolled oats, diced apple, cinnamon, and chia seeds.
2. Pour in unsweetened almond milk and stir in stevia to sweeten.
3. Seal the container and refrigerate overnight.
4. The next morning, stir the mixture and serve cold. Enjoy the classic flavors of apple pie in a nutritious, easy breakfast.

BERRY BLISS ALMOND YOGURT OVERNIGHT OATS

Ingredients
- 1/2 cup rolled oats
- 1/2 cup amond yogurt (plain, unsweetened)
- 1/2 cup mixed berries (blueberries, strawberries, raspberries)
- 1 tbsp flaxseeds
- 1 cup unsweetened almond milk
- 1 tsp stevia (adjust to taste)

Nutritional Information (per serving)
- Calories: 280
- Carbohydrates: 30g
- Proteins: 6g
- Fats: 9g

Estimated Glycemic Index: Low

Instructions:
1. In a mason jar or similar container, mix rolled oats, almond yogurt, and mixed berries.
2. Add flaxseeds for an extra nutritional boost.
3. Pour in unsweetened almond milk and mix in stevia for a touch of sweetness.
4. Seal the container and let it sit in the refrigerator overnight.
5. Stir and enjoy a creamy, berry-packed breakfast that's ready to eat right out of the fridge.

CHAPTER 4
MIDDAY DELIGHTS:

NUTRITIOUS AND SATISFYING LUNCHES

GARDEN FRESH SALAD SPECTACULAR (FOR TWO)

JACKFRUIT AVOCADO QUINOA SALAD

Ingredients
- 1 cup cooked quinoa
- 1 cup young green jackfruit (canned in brine or water, drained and shredded)
- 1 ripe avocado, diced
- 1/2 cup cherry tomatoes, halved
- 1/2 cucumber, diced
- 1/4 cup red onion, finely chopped
- Fresh cilantro, chopped (to taste)

Lemon vinaigrette:
- juice of 2 lemon
- 3 tbsp olive oil
- salt, and pepper to taste

Nutritional Information (per serving)
- Calories: 300
- Carbohydrates: 35g
- Proteins: 6g
- Fats: 18g

Estimated Glycemic Index: Low to medium

Instructions:
1. Prepare Jackfruit: If using canned jackfruit, rinse and drain it thoroughly to remove any brine. Pan-fry in a dry skillet over medium heat until slightly browned to enhance its texture and flavor.
2. Mix Salad Ingredients: In a large bowl, combine the cooked quinoa, sautéed jackfruit, diced avocado, cherry tomatoes, cucumber, and red onion.
3. Prepare Lime Vinaigrette: In a small bowl, whisk together lime juice, olive oil, salt, and pepper. Sprinkle crumbled feta cheese on top before serving.
4. Combine and Dress: Pour the lime vinaigrette over the salad ingredients and toss well to combine. Sprinkle with fresh cilantro for an added burst of flavor.
5. Serve: Enjoy this vibrant salad as a nutritious and fulfilling meal that's perfect for any occasion.

KALE & ROASTED CHICKPEA DELIGHT

Ingredients

- 2 cups chopped kale
- 1 cup roasted chickpeas (prepared from dried chickpeas, see note below)
- 1 red bell pepper, diced
- 1/4 cup sliced almonds

Tahini dressing:

- 2 tbsp tahini
- 1 lemon juiced
- 1 garlic clove minced
- water to thin
- salt to taste

Nutritional Information (per serving)

- Calories: 270
- Carbohydrates: 35g
- Proteins: 12g
- Fats: 15g

Estimated Glycemic Index: Medium

Instructions:

1. Prepare Chickpeas: Soak dried chickpeas overnight, then rinse and boil until tender. Roast the cooked chickpeas in the oven at 400°F (200°C) with a little olive oil until crispy, about 20-30 minutes. Place the chopped kale in a large salad bowl. Top with the crispy roasted chickpeas and diced red bell pepper. Add sliced almonds.
2. Tahini Dressing Preparation: In a small bowl, whisk together tahini, lemon juice, minced garlic, and a pinch of salt. Gradually add water while whisking until the dressing reaches a smooth and pourable consistency.
3. Drizzle the tahini dressing over the salad ingredients. Toss everything together to combine thoroughly. Serve this nutrient-packed meal as a flavorful and satisfying lunch or dinner.

SESAME TOFU & CRUNCHY VEGGIE BOWL

Ingredients

- 6 oz firm tofu, grilled and thinly sliced
- 2 cups mixed greens (such as spinach, arugula and romaine)
- 1/2 cup shredded carrots
- 1 tbsp sesame seeds

Lemon Ginger Dressing:

- 2 tbsp fresh lemon juice
- 1 tbsp grated ginger
- 1 tbsp olive oil
- 1 tbsp sesame oil
- salt and pepper to taste

Nutritional Information (per serving)

- Calories: 260
- Carbohydrates: 12g
- Proteins: 18g
- Fats: 17g

Estimated Glycemic Index: Low

Instructions:

1. Prepare Tofu: Grill firm tofu slices until golden brown and slightly crispy.
2. Arrange Salad: Arrange the mixed greens in a serving bowl. Drizzle the ginger soy dressing over the salad.
3. Add Ingredients: Top the greens with grilled tofu slices, shredded carrots, and sprinkle with sesame seeds.
4. Prepare Dressing: In a small bowl, whisk together fresh lemon juice, grated ginger, olive oil, sesame oil, salt, and pepper.
5. Dress Salad: Drizzle the lemon ginger dressing over the salad.
6. Toss and Serve: Toss gently to mix all the ingredients well and ensure even distribution of the dressing. Serve immediately for a fresh, protein-rich meal full of vibrant flavors and textures.

BEETROOT & WALNUT HARMONY

Ingredients

- 2 medium beetroots, roasted and sliced
- 2 cups arugula
- 1/4 cup walnut pieces
- 1/4 cup avocado, diced (for creaminess)
- 1/4 cup balsamic vinegar

Nutritional Information (per serving)

- Calories: 210
- Carbohydrates: 14g
- Proteins: 6g
- Fats: 17g

Estimated Glycemic Index: Low

Instructions:

1. Prepare Balsamic Reduction: In a small saucepan, bring 1/4 cup balsamic vinegar to a boil. Reduce heat to low and simmer until the vinegar is reduced by half and has a syrupy consistency, about 10-15 minutes. Let it cool.
2. Roast the Beetroots: Preheat the oven to 400°F (200°C). Wrap each beetroot in aluminum foil and place them on a baking sheet. Roast for 45-60 minutes or until tender when pierced with a fork. Let them cool, then peel and slice them.
3. Arrange the Salad: Lay a bed of arugula on a serving platter. Arrange the roasted beetroot slices over the arugula.
4. Add Toppings: Sprinkle walnut pieces and diced avocado evenly over the beets and arugula.
5. Dress the Salad: Drizzle with the balsamic reduction right before serving.
6. Serve: Enjoy this visually appealing and nutritious salad that's perfect for any mealtime.

BROCCOLI ALMOND BLISS SALAD

Ingredients

- Blanched broccoli florets (2 cups)
- Toasted almonds (1/4 cup)
- Dried cranberries (2 tablespoons)

Sunflower Seed Topping:

- 1/2 cup raw sunflower seeds
- 1 tablespoon lemon juice
- 1/4 teaspoon sea salt
- Water (as needed to blend)

Vegan Yogurt Dressing:

- 1/2 cup almond yogurt
- a touch of lemon juice
- 1 clove minced garlic
- Herbs of choice (such as dill or parsley)

Nutritional Information (per serving)

- Calories: 200
- Carbohydrates: 15g
- Proteins: 7g
- Fats: 15g

Estimated Glycemic Index: Medium

Instructions:

1. Prepare Broccoli: Blanch the broccoli florets in boiling water for 2 minutes, then immediately cool them in ice water to maintain their vibrant color and crisp texture.
2. Toast Almonds: In a dry skillet over medium heat, toast the almonds until golden and fragrant, stirring occasionally, about 3-5 minutes.
3. Assemble the Salad: In a large salad bowl, combine the blanched broccoli, toasted almonds, and dried cranberries.
4. Prepare Sunflower Seed Topping: Mix the ground sunflower seeds with a little lemon juice, sea salt, and water until it reaches a crumbled cheese-like consistency.
5. Make the Dressing: Whisk together the almond yogurt, lemon juice, minced garlic, and herbs until well combined and pourable
6. Dress and Toss: Sprinkle the sunflower seed topping over the salad and then drizzle with the yogurt dressing. Toss well to ensure even distribution of the dressing and topping.
7. Serve: Serve the salad chilled as a refreshing, nutrient-rich side dish or a light lunch option.

NOURISHING BOWLS FOR ENERGY-FILLED DAYS (FOR TWO)

LENTIL & AVOCADO POWER BOWL

Ingredients
- Cooked lentils (1 cup)
- Cubed avocado (1 medium avocado)
- Cherry tomatoes (1/2 cup)
- Diced cucumber (1/2 cup)

Nutritional Information (per serving)
- Calories: 320
- Carbohydrates: 30g
- Proteins: 18g
- Fats: 12g

Estimated Glycemic Index: Low

Instructions:
1. Arrange cooked lentils at the base of a bowl.
2. Top with cubed avocado, cherry tomatoes, and diced cucumber.
3. Drizzle with lemon parsley dressing.
4. Serve immediately, enjoying a fulfilling and nutritious meal perfect for energy boost.

MEDITERRANEAN HUMMUS & TOFU SALAD

Ingredients
- Mixed greens (2 cups)
- Baked tofu cubes (14 oz)
- Sliced Kalamata olives (¼ cup)
- Diced cucumber (½ cup)
- Diced tomato (½ cup)
- Olive oil (1 tbsp)

Hummus Ingredients:
- 1 cup roasted chickpeas (prepared as described in "Kale & Roasted Chickpea Delight")
- 2 tbsp tahini
- 1 clove garlic
- Juice of 1 lemon
- 2 tbsp olive oil
- Salt to taste
- Water as needed (e.g. start with ¼ cup, add more if needed)

Nutritional Information (per serving)
- Calories: 280
- Carbohydrates: 20g
- Proteins: 15g
- Fats: 18g

Estimated Glycemic Index: Low

Instructions:
1. Prepare the Baked Tofu: Preheat oven to 375°F (190°C). Drain and press 1 block of firm tofu, cut into cubes. Toss with 1 tbsp olive oil, salt, pepper, and herbs. Bake for 25-30 minutes, turning halfway, until golden and crispy.
2. Make the Hummus: Use the chickpeas prepared as per the Kale & Roasted Chickpea Delight recipe. Blend these chickpeas with tahini, minced garlic, lemon juice, olive oil, and salt in a food processor. Adjust consistency with water.
3. Lay mixed greens in a bowl. Add baked tofu cubes, spoonfuls of hummus, sliced olives, diced cucumber, and tomato. Drizzle with olive oil.

VEGAN & SAVORY BLACK BEAN WRAP

Ingredients
- Whole wheat wrap
- Seasoned tempeh or tofu, grilled and sliced (1/2 cup)
- Black beans, cooked (1/4 cup)
- Corn (1/4 cup)
- Avocado, sliced (1/4 of an avocado)

Sunflower seed cream:
- 1 cup sunflower seeds (soaked for 2-4 hours or overnight if not using a high-powered blender)
- 2 to 3 cups water (for blending, depending on desired thickness)
- Pinch of salt
- Optional: 1 tablespoon lemon juice or apple cider vinegar for a tangy flavor

Nutritional Information (per serving)
- Calories: 350
- Carbohydrates: 40g
- Proteins: 25g
- Fats: 12g

Estimated Glycemic Index: Medium

Instructions:
1. Prepare Sunflower Seeds: Drain the soaked sunflower seeds and rinse them thoroughly.
2. Blend: Place the sunflower seeds in a blender. Add the lower amount of water, salt, and optional lemon juice or vinegar. Blend on high until completely smooth. Add more water as needed to reach your desired creaminess.
3. Strain (Optional): For an ultra-smooth texture, strain the cream through a nut milk bag or fine mesh sieve
4. Lay a whole wheat wrap flat on a surface.
5. Layer grilled tempeh or tofu strips, black beans, corn, and slices of avocado. Add a dollop of sunflower seed cream.
6. Roll tightly and cut in half to serve as a hearty, balanced wrap ideal for lunch.

TROPICAL TOFU & FRUIT SALAD

Ingredients

- Mixed greens
- Grilled natural tofu (e.g., 6 oz)
- Fresh mango (1/2 mango)
- Pineapple (1/2 cup)
- Sesame seeds (1 tbsp)

Lime vinaigrette:

- 2 tbsp lime juice
- 1 tbsp olive oil

Nutritional Information (per serving)

- Calories: 250
- Carbohydrates: 15g
- Proteins: 12g
- Fats: 8g

Estimated Glycemic Index: Low

Instructions:

1. Place mixed greens in a large salad bowl.
2. Top with slices of grilled tofu, fresh mango, and pineapple chunks.
3. Sprinkle sesame seeds over the top.
4. Drizzle with lime vinaigrette and serve as a vibrant, nutrient-packed salad.

HERBED EGG & VEGGIE FIESTA

Ingredients
- Diced bell peppers (1 cup)
- Cucumbers, tomatoes (1 cup)
- Red onions, finely sliced (1/4 cup)
- Eggs (5 large)
- Fresh herbs (such as dill, basil, and oregano), chopped (1 tbsp)
- Olive oil (2 tsbp)
- Apple cider vinegar (1 tbsp)
- Kalamata olives, pitted and sliced (1/4 cup)

Nutritional Information (per serving)
- Calories: 200
- Carbohydrates: 12g
- Proteins: 10g
- Fats: 14g

Estimated Glycemic Index: Low

Instructions:
1. Prepare Eggs: Hard boil the eggs by placing them in a saucepan, covering them with water, and bringing to a boil. Once boiling, cover, turn off the heat, and let sit for 12 minutes. Cool under cold water, peel, and chop or crumble.
2. In a small bowl, combine the chopped herbs with a bit of olive oil to lightly coat the herbs, enhancing their flavors.
3. In a large salad bowl, combine diced bell peppers, cucumbers, tomatoes, and red onions. Add the chopped eggs and sliced olives to the bowl.
4. Prepare dressing: In a separate small bowl, whisk together the olive oil and apple cider vinegar to create a simple dressing.
5. Pour the dressing over the salad and gently toss to combine all the ingredients.
6. Serve this refreshing and nutritious salad as a flavorful, protein-rich meal option.

FLAVORFUL WRAPS & WHOLESOME SANDWICHES: EASY AND ENTICING LUNCHTIME SOLUTIONS (FOR TWO)

HOMEMADE WHOLE GRAIN TORTILLA RECIPE

Ingredients

- Whole wheat flour (2 cups)
- Water (3/4 cup, adjust as needed for dough consistency)
- Olive oil (2 tablespoons)
- Salt (1/2 teaspoon)

Instructions:

1. Mix Ingredients: In a large bowl, combine the whole wheat flour and salt. Add olive oil and gradually mix in water until a dough forms.
2. Knead the Dough: Turn the dough onto a floured surface and knead for about 5 minutes until smooth and elastic.
3. Rest the Dough: Cover the dough and let it rest for at least 30 minutes.
4. Divide and Roll: Divide the dough into 8 equal parts, roll each piece into a ball, and then flatten into a thin circle about 8 inches in diameter.
5. Cook the Tortillas: Preheat a large skillet over medium-high heat. Cook each tortilla for about 1 minute on each side or until the surface has brown spots and the tortilla appears cooked.

AVOCADO & QUINOA SALAD WRAP

Ingredients
- Quinoa, uncooked (1 cup)
- Avocado, sliced (1 medium)
- Cherry tomatoes, halved (½ cup)
- Fresh spinach (1 cup)
- Whole wheat tortilla (1 large)

Nutritional Information (per serving)
- Calories: 280
- Carbohydrates: 30g
- Proteins: 8g
- Fats: 14g

Estimated Glycemic Index: Low

Instructions:
1. Cook Quinoa: Rinse 1 cup of uncooked quinoa under cold water, then cook in 2 cups of water. Bring to a boil, reduce heat to low, cover, and simmer for 15-20 minutes until the water is absorbed and quinoa is tender.
2. Prepare the Quinoa Salad: Combine cooked quinoa, sliced avocado, halved cherry tomatoes, and fresh spinach leaves in a large bowl.
3. Assemble the Wrap: Spread the quinoa salad evenly over a large whole wheat tortilla.
4. Roll the Wrap: Carefully roll up the tortilla, ensuring the filling is enclosed.
5. Serve: Cut the wrap in half and serve immediately for a fresh and nutritious meal.

ZESTY BLACK BEAN AND CORN WRAP

Ingredients
- Black beans, dried (1 cup)
- Sweet corn, fresh (½ cup)
- Diced tomatoes (½ cup)
- Avocado, sliced (1 medium)
- Fresh lime juice (1 tbsp)
- Whole wheat tortilla (1 large)

Nutritional Information (per serving)
- Calories: 310
- Carbohydrates: 45g
- Proteins: 12g
- Fats: 9g

Estimated Glycemic Index: Low

Instructions:
1. Cook Black Beans: Rinse 1 cup of dried black beans then soak overnight. Drain and place in a pot, cover with fresh water, bring to a boil, then simmer for 1-2 hours until tender.
2. Prepare the Filling: Combine cooked black beans, fresh sweet corn kernels, diced tomatoes, and sliced avocado in a bowl. Add fresh lime juice and gently stir.
3. Assemble the Wrap: Spread the mixture on a large wrap.
4. Roll and Serve: Roll up the wrap tightly, cut it in half, and serve as a vibrant and hearty lunch.

CHICKPEA CAESAR WRAP

Ingredients
- Chickpeas, dried (1 cup)

Simple Caesar Dressing:
- 1/4 cup tahini (sesame seed paste)
- 1 tablespoon lemon juice
- 1 clove fresh garlic, minced
- 1 tablespoon capers, chopped
- Salt and pepper to taste
- Romaine lettuce, shredded (1 cup)
- Carrots, shredded (½ cup)
- Whole wheat tortilla (1 large)

Nutritional Information (per serving)
- Calories: 310
- Carbohydrates: 37g
- Proteins: 13g
- Fats: 10g

Estimated Glycemic Index: Low to medium

Instructions:
1. Cook Chickpeas: Rinse 1 cup of dried chickpeas then soak overnight. Drain and cook in a pot of fresh water for 1-2 hours until tender.
2. Make the Vegan Caesar Dressing: In a small bowl, whisk together tahini, lemon juice, minced garlic, and chopped capers. Add a little water if needed to reach a dressing-like consistency. Season with salt and pepper to taste.
3. Assemble the Wrap: Lay out a large tortilla. Spread the vegan Caesar dressing over the tortilla. Top with cooked chickpeas, shredded romaine lettuce, and shredded carrots.
4. Roll and Serve: Tightly roll the tortilla, slice it in half, and serve for a filling and flavorful meal.

ROASTED PEPPER AND HUMMUS WRAP

Ingredients

- Peppers, assorted colors, sliced (1 cup)
- Hummus, homemade (½ cup)
- Arugula, fresh (1 cup)
- Cucumbers, sliced (½ cup)
- Whole grain wrap (1 large)

Nutritional Information (per serving)

- Calories: 250
- Carbohydrates: 35g
- Proteins: 9g
- Fats: 6g

Estimated Glycemic Index: Low

Instructions:

1. Roast the Peppers: Preheat your oven to 425°F (220°C). Arrange sliced peppers on a baking sheet and roast for 20-25 minutes until tender and slightly charred.
2. Prepare the Hummus: Blend 1 cup of cooked chickpeas (see Chickpea Caesar Wrap for cooking instructions), 2 tablespoons of tahini, 1 clove of garlic, the juice of 1 lemon, and salt to taste in a food processor until smooth.
3. Assemble the Wrap: Spread a generous amount of hummus on a whole grain wrap. Lay roasted peppers and fresh arugula on top, add slices of cucumber.
4. Roll and Serve: Roll the wrap tightly, cut in half, and enjoy a wholesome and delicious meal.

GRILLED EGGPLANT AND FRESH PESTO PANINI

Ingredients
- Fresh eggplant slices, grilled (1 medium eggplant)
- Fresh basil pesto:
- Basil leaves (1 cup)
- Garlic (1 clove)
- pine nuts (2 tbsp)
- olive oil (2 tbsp)
- Whole grain panini bread (2 slices)

Nutritional Information (per serving)
- Calories: 230
- Carbohydrates: 26g
- Proteins: 4g
- Fats: 15g

Estimated Glycemic Index: Medium

Instructions:
1. Grill the Eggplant: Slice the eggplant about 1/2 inch thick, brush lightly with olive oil, and grill over medium heat (350°F) until tender and slightly charred, about 3-4 minutes per side.
2. Prepare Fresh Pesto: In a mortar and pestle, crush fresh basil leaves (1 cup), 1 clove of garlic, 2 tablespoons of pine nuts, and drizzle with 2 tablespoons of olive oil until smooth.
3. Assemble the Panini: Spread the fresh pesto on one side of whole grain panini bread, layer the grilled eggplant slices.
4. Grill the Panini: Heat a panini press or a grill pan over medium heat and cook the assembled panini until the bread is crispy and the cheese has melted, about 5 minutes.
5. Serve: Enjoy hot as a nutritious and fulfilling meal.

VEGAN LENTIL SALAD SANDWICH ON WHOLE GRAIN BREAD

Ingredients
- Cooked lentils (1 cup, preferably brown or green for their firmer texture)
- Diced celery (1/4 cup)
- Carrots, grated (1/4 cup)
- Red onion, finely diced (2 tbsp)
- Capers (1 tbsp, rinsed and chopped)
- Dijon mustard (1 teaspoon)
- Lemon juice (1 tbsp)
- Salt and pepper to taste
- Whole grain bread (2 slices)
- Mixed greens or romaine lettuce for garnish
- Tomato slices for garnish

Avocado Dressing:
- 1 ripe avocado
- 1 tablespoon lemon juice
- 1 clove garlic, minced
- Salt and pepper to taste
- Water to adjust consistency

Nutritional Information (per serving)
- Calories: 450
- Carbohydrates: 55g
- Proteins: 22g
- Fats: 20g

Estimated Glycemic Index: Low to medium

Instructions:
1. Prepare the Dressing: In a small bowl, mash the ripe avocado with a fork until smooth.
2. Add Flavors: Mix in the lemon juice and minced garlic. Blend well to combine. Season with salt and pepper.
3. Adjust Consistency: Depending on the desired thickness, add a little water to thin out the dressing to a spreadable consistency similar to mayonnaise.
4. Mash Lentils: In a mixing bowl, lightly mash the cooked lentils with a fork. You want them broken down slightly but retaining some texture.
5. Mix Ingredients: To the mashed lentils, add the avocado dressing, diced celery, grated carrots, chopped capers, and dijon mustard. Stir to combine well.
6. Assemble the Sandwich: Lay a bed of mixed greens or romaine lettuce on one slice of whole grain bread. Spread the lentil salad mixture over the lettuce. Top with tomato slices.
7. Complete and Serve: Place the second slice of bread on top, press gently, slice the sandwich in half, and serve immediately.

HOMEMADE VEGGIE BURGER SANDWICH

Ingredients
- Homemade veggie patty (black beans, oats, grated carrots, spices)
- Whole grain hamburger buns (1 bun)
- Lettuce, tomato, onion slices

Nutritional Information (per serving)
- Calories: 350
- Carbohydrates: 40g
- Proteins: 18g
- Fats: 9g

Estimated Glycemic Index: Medium

Instructions:
1. Prepare Veggie Patties: Combine 1 cup mashed cooked black beans, 1/2 cup finely grated carrots, 1/2 cup rolled oats, and your choice of spices (1 tsp each of cumin, paprika). Form the mixture into patties.
2. Cook the Patties: Grill or bake the patties at 375°F until they are cooked through and have a firm texture, about 25-30 minutes.
3. Assemble the Sandwich: Place a veggie patty on a whole grain bun, top with lettuce, tomato, and onion slices.
4. Serve: Enjoy this hearty and healthy alternative to traditional fast food.

CHICKPEA SALAD SANDWICH

Ingredients
- Chickpeas, cooked and mashed (1 cup)
- Sprouted grain bread (2 slices)
- Diced red onion (1/4 cup)
- Chopped celery (1/4 cup)
- Almond yogurt (1/4 cup)

Nutritional Information (per serving)
- Calories: 290
- Carbohydrates: 34g
- Proteins: 12g
- Fats: 6g

Estimated Glycemic Index: Low

Instructions:
1. Prepare Chickpea Mixture: Combine 1 cup of cooked and mashed chickpeas with diced red onion, chopped celery, and almond yogurt.
2. Assemble the Sandwich: Spread the chickpea mixture on one slice of sprouted grain bread, add a layer of fresh greens.
3. Serve: This sandwich offers a nutritious, fresh, and satisfying lunch option.

SOUL-SOOTHING SOUP SYMPHONY (FOR TWO)

HOMEMADE VEGETABLE BROTH

Ingredients

- 2 large carrots, chopped
- 2 celery stalks, chopped
- 1 onion, chopped
- 3 cloves garlic, minced
- Any other vegetable scraps (e.g., bell pepper cores, zucchini ends, mushroom stems)
- 8 cups water
- A few sprigs of fresh herbs (such as parsley, thyme, or bay leaves)
- Salt and pepper to taste

Instructions:

1. Combine all vegetables and garlic in a large pot. Add water and herbs. Bring to a boil, then reduce heat and simmer for about 1 hour.
2. Strain the broth through a fine sieve, pressing on the vegetables to extract as much liquid as possible. Season with salt and pepper to taste.
3. Use immediately or store in the refrigerator for up to 5 days, or freeze for later use.

ROASTED TOMATO BASIL SOUP

Ingredients

- Fresh tomatoes, halved (3 lbs)
- Fresh basil (1 cup)
- Garlic cloves, peeled (3 cloves)
- Olive oil (2 tbsp)
- Homemade Vegetable broth (4 cups) (from above)

Nutritional Information (per serving)

- Calories: 150
- Carbohydrates: 18g
- Proteins: 4g
- Fats: 9g

Estimated Glycemic Index: Low

Instructions:

1. Roast the Tomatoes: Preheat the oven to 400°F. Place halved tomatoes and peeled garlic cloves on a baking sheet, drizzle with olive oil, and roast for 25-30 minutes until soft and slightly charred.
2. Blend the Soup: Transfer the roasted tomatoes and garlic into a blender, add fresh basil, and puree until smooth.
3. Simmer: Pour the puree into a pot, add vegetable broth, and bring to a simmer over medium heat for about 10 minutes to blend the flavors.
4. Serve: Adjust seasoning with salt and pepper, and serve hot.

CREAMY CARROT AND GINGER SOUP

Ingredients

- Carrots, peeled and chopped (2 lbs)
- Minced fresh ginger (2 tbsp)
- Onion, chopped (1 large)
- Unsweetened Coconut milk (1 can, 14 oz)
- Homemade Vegetable broth (4 cups) (from above)

Nutritional Information (per serving)

- Calories: 140
- Carbohydrates: 22g
- Proteins: 3g
- Fats: 12g

Estimated Glycemic Index: Low

Instructions:

1. Cook the Vegetables: In a large pot, sauté the chopped onions and minced ginger in a splash of olive oil until onions are translucent. Add chopped carrots and cook for 5 minutes.
2. Simmer: Add vegetable broth to the pot and bring to a boil. Reduce heat and simmer until carrots are tender, about 20 minutes.
3. Puree and Cream: Use an immersion blender to puree the soup directly in the pot. Stir in coconut milk and heat through.
4. Serve: Serve the soup warm with a sprinkle of fresh parsley or cilantro.

SAVORY CHICKPEA AND VEGETABLE SOUP

Ingredients

- 2 tablespoons olive oil
- 1 large onion, chopped
- 2 cloves garlic, minced
- 3 medium carrots, sliced
- 3 celery stalks, sliced
- 1 cup chickpeas, soaked overnight, drained, and cooked until tender
- 2 bay leaves
- Salt and pepper, to taste
- Fresh parsley, chopped for garnish

Nutritional Information (per serving)

- Calories: 170
- Carbohydrates: 25g
- Proteins: 7g
- Fats: 3g

Estimated Glycemic Index: Low

Instructions:

1. Combine all vegetables and garlic in a large pot. Add water and herbs. Bring to a boil, then reduce heat and simmer for about 1 hour.
2. Strain the broth through a fine sieve, pressing on the vegetables to extract as much liquid as possible. Season with salt and pepper to taste.
3. Use immediately or store in the refrigerator for up to 5 days, or freeze for later use.
4. In a large pot, heat olive oil over medium heat. Add onion and garlic, and sauté until translucent. Add carrots and celery and cook until slightly tender. Pour in the homemade vegetable broth and add the cooked chickpeas and bay leaves. Bring to a boil. Reduce heat and let simmer for about 20 minutes.
5. Season with salt and pepper. Remove bay leaves. Serve hot, garnished with fresh parsley.

LENTIL AND SPINACH SOUP

Ingredients
- Dry lentils (1 cup)
- Fresh spinach, washed and trimmed
- (3 cups)
- Carrots, peeled and diced (1 cup)
- Onion, diced (1 medium)
- Garlic, minced (3 cloves)
- Homemade Vegetable broth (4 cups) (from above)

Nutritional Information (per serving)
- Calories: 230
- Carbohydrates: 30g
- Proteins: 18g
- Fats: 1g

Estimated Glycemic Index: Low

Instructions:
1. Cook Lentils: Rinse lentils thoroughly. In a large pot, combine lentils, diced carrots, diced onion, minced garlic, and vegetable broth. Bring to a boil, then reduce to a simmer.
2. Simmer: Continue to simmer until lentils are tender, about 25-30 minutes.
3. Add Spinach: Stir in fresh spinach and continue to simmer until wilted, about 3-5 minutes.
4. Serve: Season with salt and pepper, and serve the soup warm.

MUSHROOM AND BARLEY SOUP

Ingredients
- 1 tablespoon olive oil
- 1 onion, chopped
- 2 cloves garlic, minced
- 2 cups mushrooms, sliced (portobello or shiitake work well)
- 6 cups Homemade Vegetable broth (from above)
- 1 teaspoon thyme
- Salt and pepper to taste
- Chopped parsley for garnish

Nutritional Information (per serving)
- Calories: 190
- Carbohydrates: 40g
- Proteins: 9g
- Fats: 3g

Estimated Glycemic Index: Medium

Instructions:
1. In a large pot, heat olive oil over medium heat. Add onion and garlic, cooking until onion becomes translucent. Add mushrooms and cook until they start to release their juices. Stir in barley and toast lightly with the mushrooms for about 2 minutes.
2. Add vegetable broth and thyme. Bring to a boil, then reduce heat to low and cover.
3. Simmer for 30-40 minutes, or until barley is tender and most of the liquid is absorbed.
4. Season with salt and pepper. Serve garnished with parsley.

SWEET POTATO AND COCONUT SOUP

Ingredients

- Sweet potatoes, peeled and cubed (2 lbs)
- Natural Coconut milk (1 can, 14 oz)
- Onion, chopped (1 large)
- Garlic, minced (2 cloves)
- Homemade Vegetable broth (4 cups) (from above)

Nutritional Information (per serving)

- Calories: 220
- Carbohydrates: 30g
- Proteins: 3g
- Fats: 14g

Estimated Glycemic Index: Low

Instructions:

1. Cook the Sweet Potatoes: In a large pot, sauté onions and garlic in a splash of olive oil until the onions are translucent. Add the cubed sweet potatoes and cook for another 5 minutes.
2. Simmer: Add vegetable broth and bring to a boil. Reduce the heat and simmer until the sweet potatoes are soft, about 20 minutes.
3. Puree and Cream: Use an immersion blender to puree the soup in the pot. Stir in coconut milk and heat through.
4. Serve: Garnish with a sprinkle of fresh cilantro and serve warm.

SIMPLE JACKFRUIT AND WILD RICE SOUP

Ingredients

- Olive oil (1 tbs)
- Wild rice (1 cup)
- Carrots, diced (1 cup)
- Celery, diced (1 cup)
- Garlic, minced (2 cloves)
- Onion, diced (1 medium)
- Thyme, dried (1 tsp)
- Homemade Vegetable broth (6 cups) (from above)
- Young green jackfruit (fresh), rinsed and shredded
- Salt and freshly ground black pepper, to taste
- Green onions for garnish, chopped

Nutritional Information (per serving)

- Calories: 200
- Carbohydrates: 35g
- Proteins: 7g
- Fats: 3g

Estimated Glycemic Index: Low to Medium

Instructions:

1. Heat olive oil in a large pot over medium heat. Add onion and garlic; sauté until translucent.
2. Add wild rice and stir for a couple of minutes.
3. Add the homemade broth, jackfruit, carrots, celery, and thyme. Bring to a boil, then simmer for about 45-50 minutes until the rice is fully cooked.
4. Season with salt and pepper. Serve hot, garnished with green onions.

VEGAN CORN AND POTATO CHOWDER

Ingredients

- Potatoes, peeled and cubed (2 lbs)
- Corn kernels, fresh from cob (2 cups)
- Celery, diced (1 cup)
- Onion, chopped (1 medium)
- Homemade Vegetable broth (4 cups) (from above)
- Almond milk (1 cup)
- Salt and pepper to taste

Nutritional Information (per serving)

- Calories: 250
- Carbohydrates: 36g
- Proteins: 6g
- Fats: 5g

Estimated Glycemic Index: Low

Instructions:

1. Cook Potatoes and Corn: In a large pot, sauté onions and celery in a bit of olive oil until onions are translucent. Add cubed potatoes, fresh corn kernels, and vegetable broth. Bring to a boil, then reduce to a simmer until potatoes are tender, about 20 minutes.
2. Thicken the Chowder: Remove about 2 cups of the soup and blend until smooth. Return the blended mixture to the pot, stir in the amond milk, and heat through.
3. Serve: Season with salt and pepper, and serve the chowder warm, garnished with chopped chives or parsley.

GREEN GOURMET: PLANT-BASED DISHES FOR EVERY PALATE (FOR TWO)

STUFFED BELL PEPPERS WITH QUINOA AND BLACK BEANS

Ingredients
- Bell peppers (4 large, any color)
- Quinoa, uncooked (1 cup)
- Black beans, cooked (1 cup)
- Corn kernels, fresh (1 cup)
- Diced tomatoes (1 cup)
- Salt and pepper to taste

Nutritional Information (per serving)
- Calories: 250
- Carbohydrates: 45g
- Proteins: 15g
- Fats: 3g

Estimated Glycemic Index: Low

Instructions:
1. Cook Quinoa: Rinse 1 cup of quinoa under cold water. In a saucepan, cook quinoa in 2 cups of water. Bring to a boil, then simmer for 15-20 minutes until tender.
2. Prepare the Filling: Mix cooked quinoa with cooked black beans, fresh corn kernels, and diced tomatoes.
3. Prepare Bell Peppers: Slice the tops off the bell peppers, remove seeds, and rinse.
4. Stuff Peppers: Fill each pepper with the quinoa mixture and place in a baking dish.
5. Bake: Cover with foil and bake in a preheated oven at 375°F for 30 minutes. Uncover and bake for an additional 10 minutes until the peppers are tender.
6. Serve: Serve warm, topped with fresh cilantro or parsley.

VEGAN SHEPHERD'S PIE WITH LENTILS

Ingredients

- Lentils, cooked (2 cups)
- Carrots, diced (1 cup)
- Celery, diced (1 cup)
- Onion, chopped (1 medium)
- Garlic, minced (2 cloves)
- Potatoes, mashed (3 cups)
- Homemade vegetable broth (1/2 cup)
- Almond milk (1/2 cup)
- Olive oil (1 tbsp)
- Salt and pepper to taste

Nutritional Information (per serving)

- Calories: 400
- Carbohydrates: 65g
- Proteins: 18g
- Fats: 6g

Estimated Glycemic Index: Medium

Instructions:

1. Prepare Lentil Mixture: In a large skillet, sauté onions, garlic, carrots, and celery with a bit of olive oil until soft. Add cooked lentils and homemade vegetable broth. Simmer until the mixture thickens, about 10 minutes.
2. Make Mashed Potatoes: Boil potatoes until tender. Mash them with almond milk and a tablespoon of olive oil. Season with salt and pepper to taste.
3. Assemble: In a baking dish, evenly spread the lentil mixture. Top with the mashed potatoes, smoothing the surface.
4. Bake: Place in a preheated oven at 400°F for about 20 minutes, or until the top is golden and crispy.
5. Serve: Allow to cool slightly before serving. This dish can be garnished with fresh herbs like parsley for added flavor.

ZUCCHINI NOODLE PAD THAI

Ingredients
- Zucchini, spiralized (4 medium)
- Carrot, julienned (1 large)
- Red bell pepper, julienned (1 large)
- Green onions, chopped (1/4 cup)
- Peanuts, crushed (1/4 cup)

For the Pad Thai Sauce:
- Natural almond butter (2 tbsp)
- Fresh lime juice (2 tbsp)
- Crushed garlic (1 clove)
- Grated fresh ginger (1 teaspoon)
- Stevia: 1 teaspoon (adjust to taste)

Nutritional Information (per serving)
- Calories: 180
- Carbohydrates: 25g
- Proteins: 6g
- Fats: 5g

Estimated Glycemic Index: Low

Instructions:
1. Prepare Vegetables: Spiralize zucchini into noodles, julienne carrots and bell peppers.
2. Make Sauce: Combine almond butter, fresh lime juice, crushed garlic, grated ginger, and stevia in a small bowl. Whisk together until you achieve a smooth and consistent sauce.
3. Cook Pad Thai: Heat a large skillet over medium heat. Add the carrot and bell pepper, sautéing just until they start to soften. Then, add the zucchini noodles and gently toss with the sauce. Cook together for about 2 minutes, allowing the flavors to meld without overcooking the noodles.
4. Serve: Plate the Pad Thai, sprinkle with crushed peanuts and chopped green onions for added texture and flavor. Garnish with fresh cilantro and additional lime wedges if desired.

CAULIFLOWER STEAK WITH CHIMICHURRI SAUCE

Ingredients
- Cauliflower (1 large head)
- Olive oil (2 tbsp)

For Chimichurri:
- Parsley (1 cup, finely chopped)
- Garlic (3 cloves, minced)
- Olive oil (1/3 cup)
- Red wine vinegar (2 tbsp)
- Red pepper flakes (1 tbsp)
- Salt: to taste

Nutritional Information (per serving)
- Calories: 160
- Carbohydrates: 14g
- Proteins: 5g
- Fats: 12g

Estimated Glycemic Index: Low

Instructions:
1. Prepare Cauliflower Steaks: Slice cauliflower head into 1-inch thick steaks. Brush with olive oil and season with salt and pepper.
2. Grill or Roast: Grill over medium heat or roast in a preheated oven at 400°F for 25-30 minutes, flipping once, until tender and golden.
3. Make Chimichurri: Combine finely chopped parsley, minced garlic, olive oil, red wine vinegar, and a pinch of red pepper flakes in a bowl.
4. Serve: Drizzle chimichurri sauce over cauliflower steaks and serve.

MUSHROOM STROGANOFF WITH CASHEW CREAM

Ingredients
- Mixed mushrooms, sliced (2 lbs)
- Onion, chopped (1 medium)
- Garlic, minced (2 cloves)

Homemade cashew cream:
- 1 cup cashews, soaked for 4-6 hours, drained, and blended with 1/2 cup water until smooth)
- Parsley, chopped (1/4 cup)

Nutritional Information (per serving)
- Calories: 220
- Carbohydrates: 30g
- Proteins: 9g
- Fats: 14g

Estimated Glycemic Index: Low

Instructions:
1. Prepare the Cashew Cream: Soak 1 cup of cashews in water for 4-6 hours. Drain the cashews and blend with 1/2 cup fresh water until smooth. Set aside.
2. Cook Mushrooms: In a large skillet, sauté onions and garlic until translucent. Add mushrooms and cook until they release their juices and brown slightly.
3. Add Cashew Cream: Stir in cashew cream and reduce heat. Simmer for 10 minutes until the sauce thickens.
4. Serve: Sprinkle with chopped parsley and serve over cooked whole grain pasta or rice.

VEGAN JAMBALAYA WITH JACKFRUIT

Ingredients
- Fresh jackfruit, shredded (2 cups)
- Bell peppers, diced (1 cup)
- Celery, diced (1 cup)
- Onion, diced (1 large)
- Fresh tomatoes, crushed (2 cups)
- Creole seasoning (1 tbsp)

Nutritional Information (per serving)
- Calories: 200
- Carbohydrates: 45g
- Proteins: 5g
- Fats: 1g

Estimated Glycemic Index: Medium

Instructions:
1. Prepare Jackfruit: If using fresh jackfruit, remove the skin and seeds, and shred the flesh to make 2 cups.
2. Cook Vegetables: In a large pot, sauté 1 diced onion, 1 cup diced bell peppers, and 1 cup diced celery in a small amount of olive oil until soft (about 5-7 minutes).
3. Add Jackfruit and Tomatoes: Stir in the shredded jackfruit, 2 cups of freshly crushed tomatoes, and 1 tbsp of Creole seasoning. Cook for 20-30 minutes over medium heat, stirring occasionally, until the flavors meld together.
4. Serve: Serve hot, garnished with chopped green onions and a side of cooked brown rice.

SPAGHETTI SQUASH WITH TOMATO BASIL SAUCE

Ingredients
- Spaghetti squash (1 medium)
- Tomatoes, crushed (2 cups)
- Garlic, minced (2 cloves)
- Basil, fresh (1/4 cup)
- Olive oil (1 tbsp)

Nutritional Information (per serving)
- Calories: 210
- Carbohydrates: 35g
- Proteins: 4g
- Fats: 1g

Estimated Glycemic Index: Low

Instructions:
1. Roast Spaghetti Squash: Cut squash in half lengthwise, remove seeds, brush with olive oil, and season with salt and pepper. Place cut side down on a baking sheet and roast in a preheated oven at 375°F for 40-45 minutes until tender.
2. Prepare Tomato Basil Sauce: In a saucepan, sauté minced garlic in olive oil until fragrant. Add crushed tomatoes and simmer for 20 minutes. Stir in chopped basil.
3. Serve: Use a fork to shred the inside of the squash into strands. Top with tomato basil sauce and additional fresh basil.

RATATOUILLE WITH HERBED POLENTA

Ingredients
- Eggplant, sliced (1 large)
- Zucchini, sliced (2 medium)
- Bell peppers, sliced (2 large)
- Tomatoes, diced (2 large)
- Onion, sliced (1 large)
- Polenta, uncooked (1 cup)

Nutritional Information (per serving)
- Calories: 150
- Carbohydrates: 22g
- Proteins: 4g
- Fats: 3g

Estimated Glycemic Index: Low

Instructions:
1. Prepare Ratatouille: In a large skillet, layer sliced eggplant, zucchini, bell peppers, tomatoes, and onion. Drizzle with olive oil and season with salt, pepper, and mixed herbs. Cover and cook over low heat for 45-50 minutes, stirring occasionally.
2. Cook Polenta: Bring 4 cups of water to a boil. Gradually whisk in 1 cup of polenta. Reduce heat and simmer, stirring frequently, until polenta is thick and creamy, about 30 minutes.
3. Serve: Spoon ratatouille over a bed of herbed polenta and garnish with fresh basil or parsley.

CHAPTER 5
NOURISHING DINNERS: WHOLESOME EVENING MEALS

LEAN PROTEINS: FLAVORFUL AND HEALTH-CONSCIOUS CHOICES (FOR TWO)

GRILLED TOFU WITH LEMON-DILL SAUCE

Ingredients
- Tofu fillets (4 oz each)
- Lemon juice (2 tbsp)
- Fresh dill, chopped (2 tbsp)
- Unsweetened almond yogurt (1/4 cup)
- Garlic, minced (1 clove)

Nutritional Information (per serving)
- Calories: 250
- Carbohydrates: 1g
- Proteins: 25g
- Fats: 15g

Estimated Glycemic Index: Low

Instructions:
1. Prep the Tofu: Press the tofu to remove excess water, then season with salt and pepper.
2. Grill the Salmon: Preheat a grill or grill pan to medium-high heat. Brush tofu with olive oil and grill for about 3-4 minutes on each side, or until grill marks appear and tofu is heated through.
3. Make Lemon-Dill Sauce: Mix almond yogurt, lemon juice, chopped dill, and minced garlic in a bowl. Stir until smooth.
4. Serve: Drizzle lemon-dill sauce over grilled salmon and serve immediately.

HERB-ROASTED CAULIFLOWER STEAKS

Ingredients
- Cauliflower steaks (4 oz each)
- Fresh thyme (1 tbsp)
- Fresh rosemary (1 tbsp)
- Garlic, minced (2 cloves)
- Olive oil (1 tbsp)

Nutritional Information (per serving)
- Calories: 230
- Carbohydrates: 0g
- Proteins: 27g
- Fats: 5g

Estimated Glycemic Index: Low

Instructions:
1. Marinate Cauliflower: In a bowl, combine olive oil, minced garlic, chopped thyme, and rosemary. Coat the cauliflower steaks well, and let marinate for at least 30 minutes.
2. Roast Cauliflower: Preheat oven to 400°F. Place marinated cauliflower steaks on a baking sheet and roast for 20-25 minutes, until tender and golden.
3. Serve: Allow cauliflowr to rest for a few minutes before slicing. Serve hot.

SEARED TOFU STEAKS WITH AVOCADO-WASABI SAUCE

Ingredients

- Tofu steaks (4 oz each)
- Wasabi paste (1 tsp)
- Avocado (1 medium, mashed)
- A dash of salt
- Olive oil (for searing)
- Lime juice (1 tbsp)

Nutritional Information (per serving)

- Calories: 200
- Carbohydrates: 0g
- Proteins: 22g
- Fats: 9g

Estimated Glycemic Index: Low

Instructions:

1. Prepare Avocado-Wasabi Sauce: In a small bowl, mix mashed avocado with wasabi paste, a dash of salt, and lime juice until smooth.
2. Seared Tofu: Preheat a skillet over high heat and add a little olive oil. Sear tofu steaks for about 1-2 minutes per side, or until golden and crispy.
3. Serve: Slice the tofu steaks and serve them topped with avocado-wasabi sauce.

VEGAN LENTIL MEATBALLS IN TOMATO BASIL SAUCE

Ingredients

- 1 cup dried lentils, rinsed (preferably green or brown for better texture)
- 2 cups water or vegetable broth
- 1 bay leaf
- 1 tablespoon olive oil
- 1 medium onion, finely chopped
- 2 cloves garlic, minced
- 1/2 cup whole grain breadcrumbs (use gluten-free if necessary)
- 1/4 cup nutritional yeast
- 1 tablespoon flaxseed meal (mixed with 3 tablespoons water, let sit for 5 minutes to form a flax egg)
- 1 teaspoon dried oregano
- Salt and pepper to taste

For the Tomato Basil Sauce:

- 2 cups crushed tomatoes
- 2 cloves garlic, minced
- 1/2 cup fresh basil, chopped
- 1 tablespoon olive oil
- Salt and pepper to taste

Nutritional Information (per serving)

- Calories: 280
- Carbohydrates: 38g
- Proteins: 18g
- Fats: 10g

Estimated Glycemic Index: Low to medium

Instructions:

1. Cook Lentils: Bring lentils, water, and bay leaf to a boil, then simmer for 20-25 minutes until tender. Drain and remove the bay leaf.
2. Sauté Aromatics: In olive oil, sauté half the onion and garlic until translucent
3. Prepare Lentil Mixture: Pulse cooked lentils, sautéed aromatics, breadcrumbs, nutritional yeast, flax egg, oregano, salt, and pepper in a food processor.
4. Form Meatballs: Shape the mixture into 1-inch balls and place them on a baking sheet lined with parchment paper.
5. Bake Meatballs: At 375°F (190°C) for 20-25 minutes, turning halfway.
6. Make Tomato Basil Sauce: Sauté remaining garlic in olive oil, add tomatoes, and simmer. Finish with fresh basil.
7. Combine and Serve: Coat meatballs in sauce, serve over your choice of base.

QUINOA AND BLACK BEAN STUFFED BELL PEPPERS

Ingredients

- 4 large bell peppers (any color), halved and seeds removed
- 1 cup quinoa, rinsed
- 2 cups vegetable broth
- 1 can (15 ounces) black beans, drained and rinsed
- 1 cup corn kernels (fresh or frozen)
- 1 small red onion, finely chopped
- 2 cloves garlic, minced
- 1 teaspoon cumin
- 1 teaspoon chili powder
- 1/2 teaspoon paprika
- 1/2 cup fresh cilantro, chopped
- 1 lime, juiced
- 1 cup shredded vegan cheese (optional)
- Salt and pepper to taste
- Olive oil

Nutritional Information (per serving)

- Calories: 280
- Carbohydrates: 45g
- Proteins: 19g
- Fats: 8g

Estimated Glycemic Index: Low to medium

Instructions:

1. **Preheat Oven and Prepare Peppers:** Preheat your oven to 375°F (190°C). Lightly brush the outside of the bell peppers with olive oil and season the inside with salt and pepper. Set aside on a baking tray.
2. **Cook Quinoa:** In a medium saucepan, combine the quinoa and vegetable broth. Bring to a boil, cover, and reduce heat to low. Simmer for 15 minutes, or until the liquid is absorbed and the quinoa is tender.
3. **Prepare the Filling:** Heat a small amount of olive oil in a skillet over medium heat. Add the onion and garlic, sautéing until the onion is translucent. Stir in the black beans, corn, cumin, chili powder, and paprika. Cook for an additional 5 minutes, stirring occasionally. Remove from heat. Stir in the cooked quinoa, cilantro, and lime juice. Adjust seasoning with salt and pepper.
4. **Stuff the Peppers:** Spoon the quinoa and black bean mixture into each bell pepper half until well filled. Top with shredded vegan cheese if using.
5. **Bake:** Cover the tray with aluminum foil and bake for about 30 minutes. Remove the foil and continue to bake for another 10 minutes, or until the peppers are tender and the cheese is melted and bubbly.
6. **Serve:** Serve hot, garnished with additional chopped cilantro or a dollop of vegan sour cream if desired.

BAKED TOFU WITH MANGO SALSA

Ingredients

- Tofu fillets (8 oz each)
- Mango, diced (1 large)
- Red bell pepper, diced (1/2 cup)
- Red onion, finely chopped (1/4 cup)
- Cilantro, chopped (1/4 cup)
- Lime juice (2 tbsp)
- Olive oil (1 tbs)
- Salt and pepper, to taste

Nutritional Information (per serving)

- Calories: 300
- Carbohydrates: 22g
- Proteins: 15g
- Fats: 15g

Estimated Glycemic Index: Low

Instructions:

1. Prepare Mango Salsa: In a bowl, combine diced mango, diced red bell pepper, chopped red onion, chopped cilantro, and lime juice. Drizzle a little olive oil over the salsa, and season with salt and pepper to taste. Mix well and set aside to let the flavors meld.
2. Bake Tofu: Preheat your oven to 375°F (190°C). If not already pre-pressed, press the tofu to remove excess water. Slice the tofu into 1/2-inch thick steaks. Season the tofu slices with salt and pepper, and lightly coat with olive oil. Place the tofu slices on a baking sheet lined with parchment paper. Bake for 20-25 minutes until golden and slightly crispy on the edges.
3. Serve: Top the baked tofu with the fresh mango salsa and serve immediately.

VEGETABLE STIR-FRY WITH BLACK BEANS, BROCCOLI AND BELL PEPPERS

Ingredients
- Cooked black beans (1 cup)
- Broccoli, cut into florets (2 cups)
- Bell peppers, sliced (1 cup)
- Fresh ginger, grated (1 tbsp)
- Garlic, minced (2 cloves)
- Olive oil (1 tbsp)
- Salt and pepper, to taste

Nutritional Information (per serving)
- Calories: 280
- Carbohydrates: 32g
- Proteins: 12g
- Fats: 11g

Estimated Glycemic Index: Medium

Instructions:
1. Prepare Ingredients: Season the cooked black beans with salt and pepper.
2. Stir-Fry: Heat olive oil in a large skillet or wok over high heat. Add garlic and ginger, stir-fry for a few seconds to release their flavors. Add the cooked black beans and stir-fry until browned.
3. Add Vegetables: Incorporate the broccoli and bell peppers, continuing to stir-fry until the vegetables are tender but crisp, about 5-7 minutes. Adjust seasoning with additional salt and pepper if needed.
4. Serve: Serve immediately, garnished with sesame seeds if desired.

GRILLED PORTOBELLO MUSHROOMS WITH LENTIL MINT PESTO

Ingredients
- Portobello mushrooms (4 large)
- Cooked lentils (1/2 cup)
- Fresh mint leaves (1/2 cup)
- Garlic, minced (1 clove)
- Almonds, toasted (1/4 cup)
- Olive oil (3 tbsp)

Nutritional Information (per serving)
- Calories: 260
- Carbohydrates: 15g
- Proteins: 10g
- Fats: 18g

Estimated Glycemic Index: Low

Instructions:
1. Prepare Mint Lentil Pesto: In a food processor, combine cooked lentils mint leaves, toasted almonds, minced garlic, and olive oil. Blend until smooth.
2. Grill Mushrooms: Preheat grill to medium-high heat. Season Portobello mushrooms with salt and pepper, brush with olive oil, and grill for 5-7 minutes per side or until tender.
3. Serve: Serve grilled mushrooms with a dollop of lentil mint pesto.

HEALTHY SIDES: PERFECT PAIRINGS FOR A BALANCED MEAL (FOR TWO)

QUINOA SALAD WITH CUCUMBER AND AVOCADO

Ingredients
- Quinoa (1 cup cooked)
- Cucumber, diced (1 cup)
- Avocado, diced (1/4 cup)
- Lemon juice (2 tbsp)
- Olive oil (1 tbsp)
- Fresh parsley, chopped (2 tbsp)

Nutritional Information (per serving)
- Calories: 222
- Carbohydrates: 34g
- Proteins: 8g
- Fats: 7g

Estimated Glycemic Index: Low

Instructions:
1. Cook Quinoa: Rinse 1 cup of quinoa under cold water until water runs clear. Cook in 2 cups of water by bringing it to a boil, then simmering for 15 minutes. Let sit covered for 5 minutes, then fluff with a fork.
2. Prepare Salad: In a large bowl, combine cooked quinoa, diced cucumber, diced avocado, and chopped parsley.
3. Dress the Salad: Whisk together lemon juice and olive oil, drizzle over the salad, and toss to combine.
4. Serve: Chill for at least 30 minutes before serving to let flavors meld.

STEAMED ASPARAGUS WITH TOASTED ALMONDS

Ingredients
- Asparagus (1 lb)
- Almonds, slivered (1/4 cup)
- Olive oil (1 tsp)
- Salt for seasoning

Nutritional Information (per serving)
- Calories: 60
- Carbohydrates: 5g
- Proteins: 5g
- Fats: 9g

Estimated Glycemic Index: Low

Instructions:
1. Steam Asparagus: Trim the woody ends from asparagus spears. Steam in a steamer basket over boiling water for 3-4 minutes until bright green and tender-crisp.
2. Toast Almonds: In a dry skillet, toast almond slivers until golden and fragrant.
3. Serve: Toss steamed asparagus with a drizzle of olive oil and top with toasted almond slivers.

MASHED SWEET POTATOES WITH CINNAMON

Ingredients
- Sweet potatoes (2 large)
- Cinnamon (1 tsp)
- Olive oil (1 tbsp)

Nutritional Information (per serving)
- Calories: 180
- Carbohydrates: 35g
- Proteins: 2g
- Fats: 4g

Estimated Glycemic Index: Medium

Instructions:
1. Cook Sweet Potatoes: Peel and cube sweet potatoes. Boil in water until tender, about 15 minutes.
2. Mash: Drain sweet potatoes and mash with olive oil and cinnamon.
3. Serve: Serve warm, sprinkled with additional cinnamon if desired.

ROASTED BRUSSELS SPROUTS WITH BALSAMIC REDUCTION

Ingredients
- Brussels sprouts (1 lb, halved)
- Balsamic vinegar (2 tbsp)
- Olive oil (1 tbsp)

Nutritional Information (per serving)
- Calories: 150
- Carbohydrates: 20g
- Proteins: 6g
- Fats: 5g

Estimated Glycemic Index: Low

Instructions:
1. Roast Brussels Sprouts: Preheat oven to 400°F. Toss Brussels sprouts with olive oil and spread on a baking sheet. Roast for 20-25 minutes until caramelized and tender.
2. Prepare Balsamic Reduction: Simmer balsamic vinegar in a small saucepan over medium heat until thickened, about 5-7 minutes.
3. Serve: Drizzle the reduction over roasted Brussels sprouts.

CARROT AND ZUCCHINI RIBBONS WITH LEMON VINAIGRETTE

Ingredients
- Carrots, peeled into ribbons (1 cup)
- Zucchini, peeled into ribbons (1 cup)
- Lemon juice (2 tbsp)
- Olive oil (1 tbsp)
- Fresh parsley, chopped (1 tbsp)

Nutritional Information (per serving)
- Calories: 50
- Carbohydrates: 7g
- Proteins: 2g
- Fats: 5g

Estimated Glycemic Index: Low

Instructions:
1. Prepare Vegetables: Use a vegetable peeler or a mandoline to slice carrots and zucchini into thin ribbons.
2. Make Lemon Vinaigrette: In a small bowl, whisk together lemon juice, olive oil, and chopped parsley.
3. Dress and Serve: Toss the carrot and zucchini ribbons with the lemon vinaigrette. Let sit for 10 minutes to allow flavors to meld before serving.

GARLIC CAULIFLOWER "RICE"

Ingredients
- Cauliflower, grated into rice-like grains (1 head)
- Garlic, minced (2 cloves)
- Olive oil (1 tbsp)

Nutritional Information (per serving)
- Calories: 100
- Carbohydrates: 15g
- Proteins: 4g
- Fats: 7g

Estimated Glycemic Index: Low

Instructions:
1. Prepare Cauliflower Rice: Pulse cauliflower florets in a food processor until they resemble the texture of rice.
2. Cook Garlic Cauliflower Rice: Heat olive oil in a large skillet over medium heat. Add minced garlic and sauté until fragrant, about 1 minute. Add cauliflower rice and stir to coat with olive oil. Cook for 5-7 minutes, stirring frequently, until cauliflower is tender and slightly crispy.
3. Serve: Season with salt and pepper to taste and serve hot as a low-carb alternative to traditional rice.

SAUTÉED SPINACH WITH PINE NUTS AND RAISINS

Ingredients
- Spinach (1 lb)
- Pine nuts (1/4 cup)
- Raisins (1/4 cup)
- Olive oil (1 tbsp)

Nutritional Information (per serving)
- Calories: 120
- Carbohydrates: 9g
- Proteins: 5g
- Fats: 13g

Estimated Glycemic Index: Low

Instructions:
1. Toast Pine Nuts: In a dry skillet, toast pine nuts over medium heat until golden and fragrant. Remove from pan and set aside.
2. Sauté Spinach: In the same skillet, heat olive oil over medium heat. Add raisins and sauté for about 1 minute. Gradually add spinach, allowing it to wilt slightly before adding more.
3. Combine and Serve: Once all the spinach is wilted and heated through, stir in toasted pine nuts. Serve immediately, seasoned with a little salt if desired.

GRILLED CORN ON THE COB WITH CHILI LIME DRESSING

Ingredients
- Corn on the cob (4 ears)
- Lime juice (2 tbsp)
- Chili powder (1 tsp)
- Olive oil (1 tbsp)

Nutritional Information (per serving)
- Calories: 150
- Carbohydrates: 35g
- Proteins: 5g
- Fats: 5g

Estimated Glycemic Index: Low

Instructions:
1. Prepare Chili Lime Dressing: In a small bowl, mix lime juice, chili powder, and olive oil.
2. Grill Corn: Preheat grill to medium-high. Peel back corn husks and remove silk. Brush corn with some of the chili lime dressing. Grill corn, turning occasionally, until kernels are tender and slightly charred, about 10 minutes.
3. Serve: Drizzle grilled corn with remaining chili lime dressing and serve hot.

CHAPTER 6
HEALTHY INDULGENCES:

SATISFYING SNACKS FOR EVERY OCCASION

SAVORY DELIGHTS: TASTY AND WHOLESOME (FOR TWO)

SPICED CHICKPEA NUTS

Ingredients
- Chickpeas, cooked and dried (1 cup)
- Olive oil (1 tbsp)
- Cumin (1 tsp)
- Paprika (1 tsp)
- Garlic powder (1 tsp)

Nutritional Information (per serving)
- Calories: 150
- Carbohydrates: 27g
- Proteins: 8g
- Fats: 4g

Estimated Glycemic Index: Low

Instructions:
1. Prepare Chickpeas: Preheat oven to 400°F. Rinse and dry cooked chickpeas thoroughly. Toss chickpeas with olive oil, cumin, paprika, and garlic powder until evenly coated.
2. Roast: Spread chickpeas on a baking sheet in a single layer. Roast for 20-25 minutes, shaking the pan occasionally, until crispy and golden.
3. Serve: Let cool before serving. Store in an airtight container for up to a week.

ZUCCHINI CHIPS

Ingredients
- Zucchini, thinly sliced (2 medium)
- Olive oil (1 tbsp)
- Sea salt (to taste)

Nutritional Information (per serving)
- Calories: 50
- Carbohydrates: 4g
- Proteins: 2g
- Fats: 4g

Estimated Glycemic Index: Low

Instructions:
1. Prepare Zucchini: Preheat oven to 225°F. Slice zucchini very thinly using a mandoline. Toss slices with olive oil and a sprinkle of sea salt.
2. Bake: Arrange slices in a single layer on a baking sheet lined with parchment paper. Bake for 1-2 hours, flipping halfway through, until crisp and golden.
3. Serve: Let chips cool before serving. Enjoy as a light, crunchy snack.

CUCUMBER CUPS WITH HERBED ALMOND YOGURT

Ingredients
- Cucumbers (2 large, peeled and hollowed into cups)
- Unsweetened almond yogurt (1/2 cup)
- Dill, chopped (1 tbsp)
- Chives, chopped (1 tbsp)
- Garlic powder (1/2 tsp)

Nutritional Information (per serving)
- Calories: 30
- Carbohydrates: 5g
- Proteins: 6g
- Fats: 0.4g

Estimated Glycemic Index: Low

Instructions:
1. Prepare Cucumber Cups: Slice peeled cucumbers into 2-inch thick rounds. Using a melon baller or small spoon, hollow out the center of each cucumber slice to create a cup, being careful not to puncture the bottom.
2. Mix Herbed Yogurt: In a small bowl, combine almond yogurt, chopped dill, chives, and garlic powder. Mix until well combined.
3. Fill and Serve: Spoon the herbed yogurt into the cucumber cups. Chill for about 30 minutes before serving to allow the flavors to meld.

AVOCADO HUMMUS

Ingredients
- Chickpeas, cooked (1 cup)
- Avocado (1 medium)
- Homemade Tahini (2 tbsp)
- Sesame seeds (1/4 cup)
- Olive oil (1-2 tsp)
- Lemon juice (2 tbsp)
- Garlic, minced (1 clove)

Nutritional Information (per serving)
- Calories: 220
- Carbohydrates: 17g
- Proteins: 7g
- Fats: 14g

Estimated Glycemic Index: Low

Instructions:
1. Prepare Homemade Tahini: Toast sesame seeds in a dry skillet over medium heat until golden. Cool slightly. Blend in a food processor with 1-2 tsp olive oil until smooth.
2. Blend Ingredients for Hummus: In the same food processor, combine cooked chickpeas, avocado, 2 tbsp of the freshly made tahini, lemon juice, and minced garlic. Blend until smooth.
3. Adjust Consistency: If the hummus is too thick, add a little water or additional lemon juice to achieve the desired consistency.
4. Serve with a drizzle of olive oil and a sprinkle of paprika. Enjoy with fresh vegetable sticks.

BAKED KALE CHIPS

Ingredients
- Kale, leaves torn (1 bunch)
- Olive oil (1 tbsp)
- Sea salt (to taste)

Nutritional Information (per serving)
- Calories: 60
- Carbohydrates: 7g
- Proteins: 2g
- Fats: 5g

Estimated Glycemic Index: Low

Instructions:
1. **Prepare Kale**: Preheat oven to 300°F. Rinse kale and dry thoroughly. Tear leaves into bite-sized pieces.
2. **Season**: Toss kale with olive oil and a sprinkle of sea salt.
3. **Bake**: Spread kale on a baking sheet in a single layer. Bake for 10-15 minutes, until edges are crispy but not burnt.
4. **Serve**: Let cool and serve as a crispy, healthy snack.

ROASTED PUMPKIN SEEDS

Ingredients
- Pumpkin seeds (1 cup)
- Olive oil (1 tsp)
- Sea salt (1/4 tsp)

Nutritional Information (per serving)
- Calories: 180
- Carbohydrates: 3g
- Proteins: 9g
- Fats: 15g

Estimated Glycemic Index: Low

Instructions:
1. **Prepare Seeds**: Preheat oven to 350°F. Clean and dry pumpkin seeds.
2. **Season**: Toss seeds with olive oil and sea salt.
3. **Roast**: Spread seeds on a baking sheet. Roast for about 20 minutes, stirring occasionally, until golden and crispy.
4. **Serve**: Cool before serving. Store in an airtight container.

STUFFED BELL PEPPERS WITH QUINOA AND VEGETABLES

Ingredients
- Bell peppers (4 medium, tops cut off and seeded)
- Quinoa (1 cup cooked)
- Spinach, chopped (1 cup)
- Cherry tomatoes, chopped (1/2 cup)
- Onion, finely chopped (1/4 cup)
- Olive oil (1 tbsp)

Nutritional Information (per serving)
- Calories: 180
- Carbohydrates: 20g
- Proteins: 8g
- Fats: 3g

Estimated Glycemic Index: Low

Instructions:
1. Cook Quinoa: Rinse 1 cup of quinoa under cold water until water runs clear. Cook in 2 cups of water by bringing it to a boil, then simmering for 15 minutes. Let sit covered for 5 minutes, then fluff with a fork.
2. Prepare Filling: In a skillet, heat olive oil over medium heat. Sauté onion until translucent. Add chopped spinach and cherry tomatoes, cooking until spinach is wilted. Combine the vegetable mixture with the cooked quinoa. Season with salt and pepper to taste.
3. Stuff Bell Peppers: Spoon the quinoa and vegetable mixture into each hollowed-out bell pepper.
4. Bake: Place stuffed bell peppers in a baking dish. Cover with foil and bake in a preheated oven at 375°F for about 20-25 minutes until the peppers are tender.
5. Serve: Serve warm, optionally topped with a sprinkle of fresh herbs like parsley or basil.

HERB AND GARLIC MUSHROOM CAPS

Ingredients
- Large mushrooms (12 caps)
- Olive oil (2 tbsp)
- Garlic, minced (3 cloves)
- Fresh parsley, chopped (1/4 cup)

Nutritional Information (per serving)
- Calories: 90
- Carbohydrates: 10g
- Proteins: 6g
- Fats: 7g

Estimated Glycemic Index: Low

Instructions:
1. Prep Mushrooms: Preheat oven to 375°F. Remove stems from mushrooms and clean caps.
2. Season: Mix olive oil, minced garlic, and chopped parsley. Brush mixture inside and outside mushroom caps.
3. Bake: Place mushroom caps on a baking sheet. Bake for 15-20 minutes until tender.
4. Serve: Serve warm as a delicious, savory snack.

SWEET SENSATIONS: NATURALLY DELIGHTFUL (FOR TWO)

CINNAMON FLAXSEED PUDDING

Ingredients
- Ground flaxseeds (1/4 cup)
- Unsweetened almond milk (1 cup)
- Cinnamon (1 tsp)
- Vanilla extract (1/2 tsp)

Nutritional Information (per serving)
- Calories: 150
- Carbohydrates: 8g
- Proteins: 5g
- Fats: 12g

Estimated Glycemic Index: Low

Instructions:
1. Mix Ingredients: In a bowl, combine ground flaxseeds, unsweetened almond milk, cinnamon, and vanilla extract.
2. Refrigerate: Cover the bowl and refrigerate for at least 2 hours or overnight until the mixture thickens into a pudding consistency.
3. Serve: Stir well before serving. Optional: top with a few fresh berries for added flavor and nutrients.

FRESH BERRY SALAD

Ingredients
- Mixed berries (strawberries, blueberries, raspberries, 1 cup)
- Fresh mint, chopped (1 tbsp)
- Lemon zest (1 tsp)

Nutritional Information (per serving)
- Calories: 70
- Carbohydrates: 17g
- Proteins: 1g
- Fats: 0g

Estimated Glycemic Index: Low

Instructions:
1. Prepare Berries: Wash and slice strawberries if large, leaving other berries whole.
2. Mix: In a bowl, gently mix berries with chopped fresh mint and lemon zest.
3. Serve: Serve immediately or chill in the refrigerator before serving for a refreshing snack or dessert.

APPLE CINNAMON CHIPS

Ingredients
- Apples, thinly sliced (2 large)
- Cinnamon (1 tsp)

Nutritional Information (per serving)
- Calories: 95
- Carbohydrates: 25g
- Proteins: 0g
- Fats: 0g

Estimated Glycemic Index: Low

Instructions:
1. Prepare Apples: Core apples and slice thinly using a mandoline or sharp knife.
2. Season: Arrange apple slices in a single layer on a baking sheet lined with parchment paper. Sprinkle with cinnamon.
3. Bake: Bake in a preheated oven at 200°F for about 2-3 hours, flipping halfway through, until dried and crisp.
4. Serve: Cool before serving. Store in an airtight container.

CHIA AND BERRY PARFAIT

Ingredients
- Chia seeds (2 tbsp)
- Unsweetened coconut milk (1 cup)
- Mixed berries (1/2 cup)

Nutritional Information (per serving)
- Calories: 130
- Carbohydrates: 15g
- Proteins: 4g
- Fats: 9g

Estimated Glycemic Index: Low

Instructions:
1. Prepare Chia Pudding: Mix chia seeds with coconut milk. Let sit for 30 minutes or until the chia seeds have absorbed the liquid and thickened to a pudding-like consistency.
2. Assemble Parfait: In a glass, layer chia pudding alternately with mixed berries.
3. Serve: Chill in the refrigerator for at least an hour before serving.

NUTTY STUFFED PEARS

Ingredients
- Pears, halved and cored (2)
- Walnuts, chopped (1/4 cup)
- Cinnamon (1 tsp)

Nutritional Information (per serving)
- Calories: 120
- Carbohydrates: 20g
- Proteins: 2g
- Fats: 9g

Estimated Glycemic Index: Low

Instructions:
1. Prep Pears: Preheat oven to 350°F. Place halved and cored pears on a baking sheet.
2. Stuff: Mix chopped walnuts with cinnamon, and stuff into the center of each pear half.
3. Bake: Bake for 20-25 minutes until pears are soft and topping is crunchy.
4. Serve: Serve warm.

SWEET CINNAMON ALMOND MIX

Ingredients
- Almonds (1/2 cup)
- Cinnamon (1 tsp)
- Vanilla extract (1/2 tsp)
- Stevia (1/4 tsp, optional)

Nutritional Information (per serving)
- Calories: 160
- Carbohydrates: 6g
- Proteins: 6g
- Fats: 14g

Estimated Glycemic Index: Low

Instructions:
1. Prepare Nut Mix: Preheat the oven to 350°F (177°C). In a bowl, toss almonds with cinnamon, vanilla extract, and stevia if desired for a slight sweetness.
2. Roast: Spread the almond mix in a single layer on a baking sheet lined with parchment paper. Roast in the oven for 10-12 minutes, stirring halfway through, until toasted.
3. Serve: Let the almonds cool before serving. They can be stored in an airtight container for up to a week.

AVOCADO VANILLA MOUSSE

Ingredients
- Avocados (2 medium)
- Vanilla extract (1 tsp)
- Stevia (to taste)
- Unsweetened almond milk (1/4 cup)
- Fresh lemon juice (1 tbsp)

Nutritional Information (per serving)
- Calories: 240
- Carbohydrates: 12g
- Proteins: 3g
- Fats: 20g

Estimated Glycemic Index: Low

Instructions:
1. Prepare Mousse: In a food processor or blender, combine the flesh of the avocados, vanilla extract, stevia to taste, unsweetened almond milk, and fresh lemon juice. Blend until the mixture is smooth and creamy.
2. Chill: Transfer the mousse to a bowl and refrigerate for at least 1 hour to allow the flavors to meld together and the mousse to set with a thicker consistency.
3. Serve: Serve the chilled mousse in small bowls or glasses. For an extra touch of freshness, top with a few berries or a sprinkle of chopped nuts.

CARROT CAKE BALLS

Ingredients
- Carrots, grated (1 cup)
- Unsweetened dried apples, finely chopped (1/4 cup)
- Ground almonds (3/4 cup)
- Chia seeds (1/4 cup)
- Cinnamon (1/2 tsp)
- Nutmeg (a pinch)

Nutritional Information (per serving)
- Calories: 150
- Carbohydrates: 12g
- Proteins: 4g
- Fats: 9g

Estimated Glycemic Index: Low

Instructions:
1. Prepare Mixture: In a food processor, combine grated carrots, finely chopped unsweetened dried apples, ground almonds, cinnamon, and nutmeg. Process until the mixture sticks together.
2. Form Balls: Take small amounts of the mixture and roll into bite-sized balls.
3. Chill: Place the carrot cake balls on a tray and refrigerate for at least an hour to firm up.
4. Serve: Enjoy these carrot cake balls as a sweet, satisfying snack that mimics the flavors of a classic carrot cake.

GRAB-AND-GO TREATS: QUICK AND CONVENIENT (FOR TWO)

ALMOND BUTTER ENERGY BALLS

Ingredients
- Homemade Almond butter, make by blending raw almonds until smooth (1/2 cup)
- Ground flaxseeds (1/4 cup)
- Chia seeds (2 tbsp)
- Fresh Coconut, shredded: (1/4 cup)
- Stevia (to taste)

Nutritional Information (per serving)
- Calories: 150
- Carbohydrates: 8g
- Proteins: 5g
- Fats: 12g

Estimated Glycemic Index: Low

Instructions:
1. Mix Ingredients: In a bowl, combine homemade almond butter, ground flaxseeds, chia seeds, freshly shredded coconut, and stevia. Mix until well combined.
2. Form Balls: Scoop the mixture and roll into small balls, about 1-inch in diameter.
3. Chill: Place the energy balls on a baking sheet lined with parchment paper and refrigerate for at least 1 hour to set.
4. Serve: Store in an airtight container and grab a few whenever you need a quick and healthy snack.

SAVORY ROASTED CHICKPEAS

Ingredients
- Chickpeas, cooked and dried (1 cup)
- Olive oil (1 tbsp)
- Garlic powder (1 tsp)
- Paprika (1 tsp)

Nutritional Information (per serving)
- Calories: 180
- Carbohydrates: 30g
- Proteins: 9g
- Fats: 5g

Estimated Glycemic Index: Low

Instructions:
1. **Prepare Chickpeas**: Preheat oven to 400°F. Rinse and thoroughly dry cooked chickpeas. Toss chickpeas with olive oil, garlic powder, and paprika.
2. **Roast**: Spread chickpeas on a baking sheet and roast for 20-25 minutes until crispy.
3. **Serve**: Cool before serving. Store in an airtight container for a crunchy, savory snack.

CINNAMON NUT SNACK MIX

Ingredients
- Almonds (1/2 cup)
- Walnuts (1/2 cup)
- Cinnamon (1 tsp)
- Vanilla extract (1/2 tsp)

Nutritional Information (per serving)
- Calories: 160
- Carbohydrates: 6g
- Proteins: 6g
- Fats: 14g

Estimated Glycemic Index: Low

Instructions:
1. Prepare Nut Mix: Preheat the oven to 350°F (177°C). In a bowl, toss almonds and walnuts with cinnamon and vanilla extract until well coated.
2. Roast: Spread the nut mix in a single layer on a baking sheet lined with parchment paper. Roast in the oven for 10-12 minutes, stirring halfway through, until toasted.
3. Serve: Let the nuts cool before serving. Store in an airtight container for a crunchy, flavorful snack.

FLAXSEED AND BLUEBERRY PARFAIT

Ingredients
- Almond yogurt, unsweetened (1 cup)
- Blueberries (1/2 cup)
- Ground flaxseeds (2 tbsp)

Nutritional Information (per serving)
- Calories: 100
- Carbohydrates: 15g
- Proteins: 4g
- Fats: 7g

Estimated Glycemic Index: Low

Instructions:
1. Layer Ingredients: In a small jar or glass, layer Greek yogurt, fresh blueberries, and ground flaxseeds.
2. Serve: Enjoy immediately or seal with a lid and take it to go for a nutritious snack or breakfast.

PUMPKIN SEED AND SUNFLOWER SEED MIX

Ingredients
- Pumpkin seeds (1/4 cup)
- Sunflower seeds (1/4 cup)
- Sea salt (a pinch)

Nutritional Information (per serving)
- Calories: 180
- Carbohydrates: 5g
- Proteins: 9g
- Fats: 15g

Estimated Glycemic Index: Low

Instructions:
1. Mix Seeds: Combine pumpkin seeds and sunflower seeds in a bowl with a pinch of sea salt.
2. Roast (Optional): For added flavor, lightly roast the seed mix in a dry skillet over medium heat for 3-5 minutes.
3. Serve: Store in a small container or bag for an easy, portable snack.

AVOCADO AND TOMATO CUCUMBER CUPS

Ingredients
- Cucumbers (2 large, scooped into cups)
- Avocado (1 medium, mashed)
- Cherry tomatoes (1/2 cup, chopped)
- Lemon juice (1 tbsp)

Nutritional Information (per serving)
- Calories: 30
- Carbohydrates: 8g
- Proteins: 2g
- Fats: 14g

Estimated Glycemic Index: Low

Instructions:
1. Prepare Cups: Slice cucumbers into thick rounds and hollow out the centers to form cups.
2. Fill: Mix mashed avocado with chopped cherry tomatoes and lemon juice. Spoon the mixture into cucumber cups.
3. Serve: Enjoy immediately or store in a cool place for a refreshing, hydrating snack.

SPICED PEAR SLICES

Ingredients
- Pears (2 large, sliced)
- Cinnamon (1 tsp)
- Nutmeg (a pinch)

Nutritional Information (per serving)
- Calories: 100
- Carbohydrates: 27g
- Proteins: 1g
- Fats: 0g

Estimated Glycemic Index: Low

Instructions:
1. Season Pears: Toss sliced pears with cinnamon and a pinch of nutmeg.
2. Serve: Arrange on a plate and serve as a fresh, spiced snack. Perfect for an autumnal treat.

SWEET POTATO AND ALMOND BUTTER SLICES (BAKED)

Ingredients
- Sweet potatoes (2 medium, sliced)
- Homemade Almond butter make by blending raw almonds until smooth (1/4 cup)

Nutritional Information (per serving)
- Calories: 112
- Carbohydrates: 20g
- Proteins: 3g
- Fats: 6g

Estimated Glycemic Index: Low

Instructions:
1. Prepare Sweet Potato Slices: Preheat the oven to 375°F (190°C). Slice sweet potatoes into 1/4 inch thick rounds.
2. Bake: Arrange the sweet potato slices in a single layer on a baking sheet lined with parchment paper. Bake in the preheated oven for about 25-30 minutes, or until the slices are tender.
3. Top and Serve: Remove the sweet potato slices from the oven and allow them to cool slightly. Spread a thin layer of almond butter on each slice. Serve warm or at room temperature for a satisfying snack.

CHAPTER 7
EXCLUSIVE BONUSES

Unlock Exclusive Bonuses with QR Code!

1. Hydration and Drinks: Secrets by Barbara O'Neill

Discover the ultimate guide to staying hydrated and nourishing your body with delicious, natural drinks. Barbara O'Neill shares her secrets on how to prepare the best beverages for optimal health and hydration.

2. "Ayurveda for Women": The Ultimate Wellness Guide

Dive into the incredible book "Ayurveda for Women." Learn how to align your diet with the natural rhythms of your body. This comprehensive guide teaches you how to enhance your health and well-being through natural, Ayurvedic nutrition.

3. Healing with Nature: Barbara O'Neill's Natural Remedies

Discover the Art of Home Remedies with Barbara O'Neill's teachings on natural healing. This booklet provides invaluable insights into using everyday ingredients to treat and maintain health naturally.

4. Living Well with Diabetes: Exercise and Stress Reduction

Gain access to expert advice on managing diabetes through exercise and stress reduction. Barbara O'Neill provides practical solutions to help you live a balanced, healthy life despite diabetes.

5. Cure for High Blood Pressure: Inspired by Barbara O'Neill

Discover natural methods to manage and reduce high blood pressure with tips inspired by Barbara O'Neill. This bonus covers lifestyle changes and natural remedies to keep your blood pressure in check.

CHAPTER 8
6-MONTH MEAL PLAN

WEEK 1

Day 1:
- Breakfast: Velvety Blueberry Spinach Bliss Smoothie
- Lunch: Jackfruit Avocado Quinoa Salad
- Snack: Cucumber Cups with Herbed Almond Yogurt
- Dinner: Grilled Tofu with Lemon-Dill Sauce
  - Side: Steamed Asparagus with Toasted Almonds

Day 2:
- Breakfast: Crunchy Seed & Nut Morning Muesli
- Lunch: Kale & Roasted Chickpea Delight
- Snack: Zucchini Chips
- Dinner: Herb-Roasted Cauliflower Steaks
  - Side: Garlic Cauliflower "Rice"

Day 3:
- Breakfast: Warm Chia & Golden Hemp Heart Porridge
- Lunch: Beetroot & Walnut Harmony
- Snack: Fresh Berry Salad
- Dinner: Seared Tofu Steaks with Avocado-Wasabi Sauce
  - Side: Carrot and Zucchini Ribbons with Lemon Vinaigrette

Day 4:
- Breakfast: Apple Cinnamon Delight Shake
- Lunch: Sesame Tofu & Crunchy Veggie Bowl
- Snack: Apple Cinnamon Chips
- Dinner: Vegan Lentil Meatballs in Tomato Basil Sauce
  - Side: Quinoa Salad with Cucumber and Avocado

Day 5:
- Breakfast: Mango & Coconut Tropical Escape Smoothie
- Lunch: Mediterranean Hummus & Tofu Salad
- Snack: Baked Kale Chips
- Dinner: Baked Tofu with Mango Salsa
  - Side: Mashed Sweet Potatoes with Cinnamon

Day 6:
- Breakfast: Energizing Espresso & Oat Smoothie

- Lunch: Herbed Egg & Veggie Fiesta
- Snack: Spiced Chickpea Nuts
- Dinner: Vegetable Stir-Fry with Black Beans, Broccoli and Bell Peppers
  - Side: Roasted Brussels Sprouts with Balsamic Reduction

Day 7:
- Breakfast: Heart-Healthy Avocado & Mixed Berry Smoothie
- Lunch: Garden Fresh Veggie & Egg White Frittata
- Snack: Nutty Stuffed Pears
- Dinner: Grilled Portobello Mushrooms with Lentil Mint Pesto
  - Side: Sautéed Spinach with Pine Nuts and Raisins

WEEK 2

Day 8:
- Breakfast: Heart-Healthy Avocado & Mixed Berry Smoothie
- Lunch: Quinoa Power Breakfast Bowl
- Snack: Flaxseed and Blueberry Parfait
- Dinner: Cauliflower Steak with Chimichurri Sauce
  - Side: Steamed Asparagus with Toasted Almonds

Day 9:
- Breakfast: Cinnamon Spiced Quinoa Breakfast Cereal
- Lunch: Kale & Roasted Chickpea Delight
- Snack: Roasted Pumpkin Seeds
- Dinner: Vegan Shepherd's Pie with Lentils
  - Side: Garlic Cauliflower "Rice"

Day 10:
- Breakfast: Buckwheat & Toasted Almond Morning Bowl
- Lunch: Garden Fresh Veggie & Egg White Frittata
- Snack: Avocado Hummus
- Dinner: Cauliflower Steak with Chimichurri Sauce
  - Side: Carrot and Zucchini Ribbons with Lemon Vinaigrette

Day 11:
- Breakfast: Mango & Coconut Tropical Escape Smoothie
- Lunch: Beetroot & Walnut Harmony
- Snack: Sweet Cinnamon Almond Mix
- Dinner: Vegan Lentil Meatballs in Tomato Basil Sauce
  - Side: Roasted Brussels Sprouts with Balsamic Reduction

Day 12:
- Breakfast: Summer Berries & Amaranth Cereal Bowl
- Lunch: Mediterranean Hummus & Tofu Salad
- Snack: Spiced Pear Slices

- Dinner: Seared Tofu Steaks with Avocado-Wasabi Sauce
    - Side: Quinoa Salad with Cucumber and Avocado

Day 13:
- Breakfast: Warm Chia & Golden Hemp Heart Porridge
- Lunch: Kale & Roasted Chickpea Delight
- Snack: Apple Cinnamon Chips
- Dinner: Baked Tofu with Mango Salsa
    - Side: Mashed Sweet Potatoes with Cinnamon

Day 14:
- Breakfast: Crunchy Seed & Nut Morning Muesli
- Lunch: Broccoli Almond Bliss Salad
- Snack: Baked Kale Chips
- Dinner: Vegetable Stir-Fry with Black Beans, Broccoli and Bell Peppers
    - Side: Sautéed Spinach with Pine Nuts and Raisins

WEEK 3

Day 15:
- Breakfast: Cinnamon Spiced Quinoa Breakfast Cereal
- Lunch: Beetroot & Walnut Harmony
- Snack: Avocado Hummus
- Dinner: Vegan Shepherd's Pie with Lentils
    - Side: Steamed Asparagus with Toasted Almonds

Day 16:
- Breakfast: Buckwheat & Toasted Almond Morning Bowl
- Lunch: Garden Fresh Veggie & Egg White Frittata
- Snack: Roasted Pumpkin Seeds
- Dinner: Cauliflower Steak with Chimichurri Sauce
    - Side: Garlic Cauliflower "Rice"

Day 17:
- Breakfast: Mango & Coconut Tropical Escape Smoothie
- Lunch: Sesame Tofu & Crunchy Veggie Bowl
- Snack: Sweet Cinnamon Almond Mix
- Dinner: Vegan Lentil Meatballs in Tomato Basil Sauce
    - Side: Carrot and Zucchini Ribbons with Lemon Vinaigrette

Day 18:
- Breakfast: Summer Berries & Amaranth Cereal Bowl
- Lunch: Mediterranean Hummus & Tofu Salad
- Snack: Baked Kale Chips
- Dinner: Seared Tofu Steaks with Avocado-Wasabi Sauce
    - Side: Quinoa Salad with Cucumber and Avocado

Day 19:
- Breakfast: Warm Chia & Golden Hemp Heart Porridge
- Lunch: Kale & Roasted Chickpea Delight
- Snack: Zucchini Chips
- Dinner: Baked Tofu with Mango Salsa
  - Side: Mashed Sweet Potatoes with Cinnamon

Day 20:
- Breakfast: Crunchy Seed & Nut Morning Muesli
- Lunch: Herbed Egg & Veggie Fiesta
- Snack: Fresh Berry Salad
- Dinner: Vegetable Stir-Fry with Black Beans, Broccoli and Bell Peppers
  - Side: Roasted Brussels Sprouts with Balsamic Reduction

Day 21:
- Breakfast: Energizing Espresso & Oat Smoothie
- Lunch: Quinoa Power Breakfast Bowl
- Snack: Apple Cinnamon Chips
- Dinner: Grilled Portobello Mushrooms with Lentil Mint Pesto
  - Side: Sautéed Spinach with Pine Nuts and Raisins

WEEK 4

Day 22:
- Breakfast: Heart-Healthy Avocado & Mixed Berry Smoothie
- Lunch: Jackfruit Avocado Quinoa Salad
- Snack: Cucumber Cups with Herbed Almond Yogurt
- Dinner: Herb-Roasted Cauliflower Steaks
  - Side: Garlic Cauliflower "Rice"

Day 23:
- Breakfast: Velvety Blueberry Spinach Bliss Smoothie
- Lunch: Kale & Roasted Chickpea Delight
- Snack: Spiced Chickpea Nuts
- Dinner: Seared Tofu Steaks with Avocado-Wasabi Sauce
  - Side: Steamed Asparagus with Toasted Almonds

Day 24:
- Breakfast: Warm Chia & Golden Hemp Heart Porridge
- Lunch: Beetroot & Walnut Harmony
- Snack: Baked Kale Chips
- Dinner: Vegan Lentil Meatballs in Tomato Basil Sauce
  - Side: Quinoa Salad with Cucumber and Avocado

Day 25:
- Breakfast: Mango & Coconut Tropical Escape Smoothie

- Lunch: Sesame Tofu & Crunchy Veggie Bowl
- Snack: Nutty Stuffed Pears
- Dinner: Cauliflower Steak with Chimichurri Sauce
  - Side: Carrot and Zucchini Ribbons with Lemon Vinaigrette

Day 26:
- Breakfast: Apple Cinnamon Delight Shake
- Lunch: Mediterranean Hummus & Tofu Salad
- Snack: Roasted Pumpkin Seeds
- Dinner: Baked Tofu with Mango Salsa
  - Side: Mashed Sweet Potatoes with Cinnamon

Day 27:
- Breakfast: Energizing Espresso & Oat Smoothie
- Lunch: Herbed Egg & Veggie Fiesta
- Snack: Zucchini Chips
- Dinner: Vegetable Stir-Fry with Black Beans, Broccoli and Bell Peppers
  - Side: Roasted Brussels Sprouts with Balsamic Reduction

Day 28:
- Breakfast: Energizing Espresso & Oat Smoothie
- Lunch: Garden Fresh Veggie & Egg White Frittata
- Snack: Fresh Berry Salad
- Dinner: Grilled Portobello Mushrooms with Lentil Mint Pesto
  - Side: Sautéed Spinach with Pine Nuts and Raisins

WEEK 5

Day 29:
- Breakfast: Cinnamon Spiced Quinoa Breakfast Cereal
- Lunch: Broccoli Almond Bliss Salad
- Snack: Avocado Hummus
- Dinner: Vegan Shepherd's Pie with Lentils
  - Side: Steamed Asparagus with Toasted Almonds

Day 30:
- Breakfast: Buckwheat & Toasted Almond Morning Bowl
- Lunch: Garden Fresh Veggie & Egg White Frittata
- Snack: Roasted Pumpkin Seeds
- Dinner: Cauliflower Steak with Chimichurri Sauce
  - Side: Garlic Cauliflower "Rice"

Day 31:
- Breakfast: Mango & Coconut Tropical Escape Smoothie
- Lunch: Sesame Tofu & Crunchy Veggie Bowl
- Snack: Sweet Cinnamon Almond Mix

- Dinner: Vegan Lentil Meatballs in Tomato Basil Sauce
  - Side: Carrot and Zucchini Ribbons with Lemon Vinaigrette

Day 32:
- Breakfast: Summer Berries & Amaranth Cereal Bowl
- Lunch: Mediterranean Hummus & Tofu Salad
- Snack: Baked Kale Chips
- Dinner: Seared Tofu Steaks with Avocado-Wasabi Sauce
  - Side: Quinoa Salad with Cucumber and Avocado

Day 33:
- Breakfast: Warm Chia & Golden Hemp Heart Porridge
- Lunch: Kale & Roasted Chickpea Delight
- Snack: Zucchini Chips
- Dinner: Baked Tofu with Mango Salsa
  - Side: Mashed Sweet Potatoes with Cinnamon

Day 34:
- Breakfast: Crunchy Seed & Nut Morning Muesli
- Lunch: Herbed Egg & Veggie Fiesta
- Snack: Fresh Berry Salad
- Dinner: Vegetable Stir-Fry with Black Beans, Broccoli and Bell Peppers
  - Side: Roasted Brussels Sprouts with Balsamic Reduction

Day 35:
- Breakfast: Energizing Espresso & Oat Smoothie
- Lunch: Quinoa Power Breakfast Bowl
- Snack: Apple Cinnamon Chips
- Dinner: Grilled Portobello Mushrooms with Lentil Mint Pesto
  - Side: Sautéed Spinach with Pine Nuts and Raisins

WEEK 6

Day 36:
- Breakfast: Heart-Healthy Avocado & Mixed Berry Smoothie
- Lunch: Jackfruit Avocado Quinoa Salad
- Snack: Cucumber Cups with Herbed Almond Yogurt
- Dinner: Herb-Roasted Cauliflower Steaks
  - Side: Garlic Cauliflower "Rice"

Day 37:
- Breakfast: Velvety Blueberry Spinach Bliss Smoothie
- Lunch: Kale & Roasted Chickpea Delight
- Snack: Spiced Chickpea Nuts
- Dinner: Seared Tofu Steaks with Avocado-Wasabi Sauce
  - Side: Steamed Asparagus with Toasted Almonds

Day 38:
- Breakfast: Warm Chia & Golden Hemp Heart Porridge
- Lunch: Beetroot & Walnut Harmony
- Snack: Baked Kale Chips
- Dinner: Vegan Lentil Meatballs in Tomato Basil Sauce
  - Side: Quinoa Salad with Cucumber and Avocado

Day 39:
- Breakfast: Mango & Coconut Tropical Escape Smoothie
- Lunch: Sesame Tofu & Crunchy Veggie Bowl
- Snack: Nutty Stuffed Pears
- Dinner: Cauliflower Steak with Chimichurri Sauce
  - Side: Carrot and Zucchini Ribbons with Lemon Vinaigrette

Day 40:
- Breakfast: Apple Cinnamon Delight Shake
- Lunch: Mediterranean Hummus & Tofu Salad
- Snack: Roasted Pumpkin Seeds
- Dinner: Baked Tofu with Mango Salsa
  - Side: Mashed Sweet Potatoes with Cinnamon

Day 41:
- Breakfast: Energizing Espresso & Oat Smoothie
- Lunch: Herbed Egg & Veggie Fiesta
- Snack: Zucchini Chips
- Dinner: Vegetable Stir-Fry with Black Beans, Broccoli and Bell Peppers
  - Side: Roasted Brussels Sprouts with Balsamic Reduction

Day 42:
- Breakfast: Energizing Espresso & Oat Smoothie
- Lunch: Garden Fresh Veggie & Egg White Frittata
- Snack: Fresh Berry Salad
- Dinner: Grilled Portobello Mushrooms with Lentil Mint Pesto
  - Side: Sautéed Spinach with Pine Nuts and Raisins

WEEK 7

Day 43:
- Breakfast: Cinnamon Spiced Quinoa Breakfast Cereal
- Lunch: Broccoli Almond Bliss Salad
- Snack: Avocado Hummus
- Dinner: Vegan Shepherd's Pie with Lentils
  - Side: Steamed Asparagus with Toasted Almonds

Day 44:
- Breakfast: Buckwheat & Toasted Almond Morning Bowl

- Lunch: Garden Fresh Veggie & Egg White Frittata
- Snack: Roasted Pumpkin Seeds
- Dinner: Cauliflower Steak with Chimichurri Sauce
  - Side: Garlic Cauliflower "Rice"

Day 45:
- Breakfast: Mango & Coconut Tropical Escape Smoothie
- Lunch: Sesame Tofu & Crunchy Veggie Bowl
- Snack: Sweet Cinnamon Almond Mix
- Dinner: Vegan Lentil Meatballs in Tomato Basil Sauce
  - Side: Carrot and Zucchini Ribbons with Lemon Vinaigrette

Day 46:
- Breakfast: Summer Berries & Amaranth Cereal Bowl
- Lunch: Mediterranean Hummus & Tofu Salad
- Snack: Baked Kale Chips
- Dinner: Seared Tofu Steaks with Avocado-Wasabi Sauce
  - Side: Quinoa Salad with Cucumber and Avocado

Day 47:
- Breakfast: Warm Chia & Golden Hemp Heart Porridge
- Lunch: Kale & Roasted Chickpea Delight
- Snack: Zucchini Chips
- Dinner: Baked Tofu with Mango Salsa
  - Side: Mashed Sweet Potatoes with Cinnamon

Day 48:
- Breakfast: Crunchy Seed & Nut Morning Muesli
- Lunch: Herbed Egg & Veggie Fiesta
- Snack: Fresh Berry Salad
- Dinner: Vegetable Stir-Fry with Black Beans, Broccoli and Bell Peppers
  - Side: Roasted Brussels Sprouts with Balsamic Reduction

Day 49:
- Breakfast: Energizing Espresso & Oat Smoothie
- Lunch: Quinoa Power Breakfast Bowl
- Snack: Apple Cinnamon Chips
- Dinner: Grilled Portobello Mushrooms with Lentil Mint Pesto
  - Side: Sautéed Spinach with Pine Nuts and Raisins

WEEK 8

Day 50:
- Breakfast: Heart-Healthy Avocado & Mixed Berry Smoothie
- Lunch: Jackfruit Avocado Quinoa Salad
- Snack: Cucumber Cups with Herbed Almond Yogurt

- Dinner: Herb-Roasted Cauliflower Steaks
  - Side: Garlic Cauliflower "Rice"

Day 51:
- Breakfast: Velvety Blueberry Spinach Bliss Smoothie
- Lunch: Kale & Roasted Chickpea Delight
- Snack: Spiced Chickpea Nuts
- Dinner: Seared Tofu Steaks with Avocado-Wasabi Sauce
  - Side: Steamed Asparagus with Toasted Almonds

Day 52:
- Breakfast: Warm Chia & Golden Hemp Heart Porridge
- Lunch: Beetroot & Walnut Harmony
- Snack: Baked Kale Chips
- Dinner: Vegan Lentil Meatballs in Tomato Basil Sauce
  - Side: Quinoa Salad with Cucumber and Avocado

Day 53:
- Breakfast: Mango & Coconut Tropical Escape Smoothie
- Lunch: Sesame Tofu & Crunchy Veggie Bowl
- Snack: Nutty Stuffed Pears
- Dinner: Cauliflower Steak with Chimichurri Sauce
  - Side: Carrot and Zucchini Ribbons with Lemon Vinaigrette

Day 54:
- Breakfast: Apple Cinnamon Delight Shake
- Lunch: Mediterranean Hummus & Tofu Salad
- Snack: Roasted Pumpkin Seeds
- Dinner: Baked Tofu with Mango Salsa
  - Side: Mashed Sweet Potatoes with Cinnamon

Day 55:
- Breakfast: Energizing Espresso & Oat Smoothie
- Lunch: Herbed Egg & Veggie Fiesta
- Snack: Zucchini Chips
- Dinner: Vegetable Stir-Fry with Black Beans, Broccoli and Bell Peppers
  - Side: Roasted Brussels Sprouts with Balsamic Reduction

Day 56:
- Breakfast: Energizing Espresso & Oat Smoothie
- Lunch: Garden Fresh Veggie & Egg White Frittata
- Snack: Fresh Berry Salad
- Dinner: Grilled Portobello Mushrooms with Lentil Mint Pesto
  - Side: Sautéed Spinach with Pine Nuts and Raisins

WEEK 9

Day 57:
- Breakfast: Cinnamon Spiced Quinoa Breakfast Cereal
- Lunch: Broccoli Almond Bliss Salad
- Snack: Avocado Hummus
- Dinner: Vegan Shepherd's Pie with Lentils
  - Side: Steamed Asparagus with Toasted Almonds

Day 58:
- Breakfast: Buckwheat & Toasted Almond Morning Bowl
- Lunch: Garden Fresh Veggie & Egg White Frittata
- Snack: Roasted Pumpkin Seeds
- Dinner: Cauliflower Steak with Chimichurri Sauce
  - Side: Garlic Cauliflower "Rice"

Day 59:
- Breakfast: Mango & Coconut Tropical Escape Smoothie
- Lunch: Sesame Tofu & Crunchy Veggie Bowl
- Snack: Sweet Cinnamon Almond Mix
- Dinner: Vegan Lentil Meatballs in Tomato Basil Sauce
  - Side: Carrot and Zucchini Ribbons with Lemon Vinaigrette

Day 60:
- Breakfast: Summer Berries & Amaranth Cereal Bowl
- Lunch: Mediterranean Hummus & Tofu Salad
- Snack: Baked Kale Chips
- Dinner: Seared Tofu Steaks with Avocado-Wasabi Sauce
  - Side: Quinoa Salad with Cucumber and Avocado

Day 61:
- Breakfast: Warm Chia & Golden Hemp Heart Porridge
- Lunch: Kale & Roasted Chickpea Delight
- Snack: Zucchini Chips
- Dinner: Baked Tofu with Mango Salsa
  - Side: Mashed Sweet Potatoes with Cinnamon

Day 62:
- Breakfast: Crunchy Seed & Nut Morning Muesli
- Lunch: Herbed Egg & Veggie Fiesta
- Snack: Fresh Berry Salad
- Dinner: Vegetable Stir-Fry with Black Beans, Broccoli and Bell Peppers
  - Side: Roasted Brussels Sprouts with Balsamic Reduction

Day 63:
- Breakfast: Energizing Espresso & Oat Smoothie

- Lunch: Quinoa Power Breakfast Bowl
- Snack: Apple Cinnamon Chips
- Dinner: Grilled Portobello Mushrooms with Lentil Mint Pesto
  - Side: Sautéed Spinach with Pine Nuts and Raisins

WEEK 10

Day 64:
- Breakfast: Heart-Healthy Avocado & Mixed Berry Smoothie
- Lunch: Jackfruit Avocado Quinoa Salad
- Snack: Cucumber Cups with Herbed Almond Yogurt
- Dinner: Herb-Roasted Cauliflower Steaks
  - Side: Garlic Cauliflower "Rice"

Day 65:
- Breakfast: Velvety Blueberry Spinach Bliss Smoothie
- Lunch: Kale & Roasted Chickpea Delight
- Snack: Spiced Chickpea Nuts
- Dinner: Seared Tofu Steaks with Avocado-Wasabi Sauce
  - Side: Steamed Asparagus with Toasted Almonds

Day 66:
- Breakfast: Warm Chia & Golden Hemp Heart Porridge
- Lunch: Beetroot & Walnut Harmony
- Snack: Baked Kale Chips
- Dinner: Vegan Lentil Meatballs in Tomato Basil Sauce
  - Side: Quinoa Salad with Cucumber and Avocado

Day 67:
- Breakfast: Mango & Coconut Tropical Escape Smoothie
- Lunch: Sesame Tofu & Crunchy Veggie Bowl
- Snack: Nutty Stuffed Pears
- Dinner: Cauliflower Steak with Chimichurri Sauce
  - Side: Carrot and Zucchini Ribbons with Lemon Vinaigrette

Day 68:
- Breakfast: Apple Cinnamon Delight Shake
- Lunch: Mediterranean Hummus & Tofu Salad
- Snack: Roasted Pumpkin Seeds
- Dinner: Baked Tofu with Mango Salsa
  - Side: Mashed Sweet Potatoes with Cinnamon

Day 69:
- Breakfast: Energizing Espresso & Oat Smoothie
- Lunch: Herbed Egg & Veggie Fiesta
- Snack: Zucchini Chips

- Dinner: Vegetable Stir-Fry with Black Beans, Broccoli and Bell Peppers
  - Side: Roasted Brussels Sprouts with Balsamic Reduction

Day 70:
- Breakfast: Energizing Espresso & Oat Smoothie
- Lunch: Garden Fresh Veggie & Egg White Frittata
- Snack: Fresh Berry Salad
- Dinner: Grilled Portobello Mushrooms with Lentil Mint Pesto
  - Side: Sautéed Spinach with Pine Nuts and Raisins

WEEK 11

Day 71:
- Breakfast: Cinnamon Spiced Quinoa Breakfast Cereal
- Lunch: Broccoli Almond Bliss Salad
- Snack: Avocado Hummus
- Dinner: Vegan Shepherd's Pie with Lentils
  - Side: Steamed Asparagus with Toasted Almonds

Day 72:
- Breakfast: Buckwheat & Toasted Almond Morning Bowl
- Lunch: Garden Fresh Veggie & Egg White Frittata
- Snack: Roasted Pumpkin Seeds
- Dinner: Cauliflower Steak with Chimichurri Sauce
  - Side: Garlic Cauliflower "Rice"

Day 73:
- Breakfast: Mango & Coconut Tropical Escape Smoothie
- Lunch: Sesame Tofu & Crunchy Veggie Bowl
- Snack: Sweet Cinnamon Almond Mix
- Dinner: Vegan Lentil Meatballs in Tomato Basil Sauce
  - Side: Carrot and Zucchini Ribbons with Lemon Vinaigrette

Day 74:
- Breakfast: Summer Berries & Amaranth Cereal Bowl
- Lunch: Mediterranean Hummus & Tofu Salad
- Snack: Baked Kale Chips
- Dinner: Seared Tofu Steaks with Avocado-Wasabi Sauce
  - Side: Quinoa Salad with Cucumber and Avocado

Day 75:
- Breakfast: Warm Chia & Golden Hemp Heart Porridge
- Lunch: Kale & Roasted Chickpea Delight
- Snack: Zucchini Chips
- Dinner: Baked Tofu with Mango Salsa
  - Side: Mashed Sweet Potatoes with Cinnamon

Day 76:
- Breakfast: Crunchy Seed & Nut Morning Muesli
- Lunch: Herbed Egg & Veggie Fiesta
- Snack: Fresh Berry Salad
- Dinner: Vegetable Stir-Fry with Black Beans, Broccoli and Bell Peppers
  - Side: Roasted Brussels Sprouts with Balsamic Reduction

Day 77:
- Breakfast: Energizing Espresso & Oat Smoothie
- Lunch: Quinoa Power Breakfast Bowl
- Snack: Apple Cinnamon Chips
- Dinner: Grilled Portobello Mushrooms with Lentil Mint Pesto
  - Side: Sautéed Spinach with Pine Nuts and Raisins

WEEK 12

Day 78:
- Breakfast: Heart-Healthy Avocado & Mixed Berry Smoothie
- Lunch: Jackfruit Avocado Quinoa Salad
- Snack: Cucumber Cups with Herbed Almond Yogurt
- Dinner: Herb-Roasted Cauliflower Steaks
  - Side: Garlic Cauliflower "Rice"

Day 79:
- Breakfast: Velvety Blueberry Spinach Bliss Smoothie
- Lunch: Kale & Roasted Chickpea Delight
- Snack: Spiced Chickpea Nuts
- Dinner: Seared Tofu Steaks with Avocado-Wasabi Sauce
  - Side: Steamed Asparagus with Toasted Almonds

Day 80:
- Breakfast: Warm Chia & Golden Hemp Heart Porridge
- Lunch: Beetroot & Walnut Harmony
- Snack: Baked Kale Chips
- Dinner: Vegan Lentil Meatballs in Tomato Basil Sauce
  - Side: Quinoa Salad with Cucumber and Avocado

Day 81:
- Breakfast: Mango & Coconut Tropical Escape Smoothie
- Lunch: Sesame Tofu & Crunchy Veggie Bowl
- Snack: Nutty Stuffed Pears
- Dinner: Cauliflower Steak with Chimichurri Sauce
  - Side: Carrot and Zucchini Ribbons with Lemon Vinaigrette

Day 82:
- Breakfast: Apple Cinnamon Delight Shake

- Lunch: Mediterranean Hummus & Tofu Salad
- Snack: Roasted Pumpkin Seeds
- Dinner: Baked Tofu with Mango Salsa
  - Side: Mashed Sweet Potatoes with Cinnamon

Day 83:
- Breakfast: Energizing Espresso & Oat Smoothie
- Lunch: Herbed Egg & Veggie Fiesta
- Snack: Zucchini Chips
- Dinner: Vegetable Stir-Fry with Black Beans, Broccoli and Bell Peppers
  - Side: Roasted Brussels Sprouts with Balsamic Reduction

Day 84:
- Breakfast: Energizing Espresso & Oat Smoothie
- Lunch: Garden Fresh Veggie & Egg White Frittata
- Snack: Fresh Berry Salad
- Dinner: Grilled Portobello Mushrooms with Lentil Mint Pesto
  - Side: Sautéed Spinach with Pine Nuts and Raisins

WEEK 13

Day 85:
- Breakfast: Cinnamon Spiced Quinoa Breakfast Cereal
- Lunch: Broccoli Almond Bliss Salad
- Snack: Avocado Hummus
- Dinner: Vegan Shepherd's Pie with Lentils
  - Side: Steamed Asparagus with Toasted Almonds

Day 86:
- Breakfast: Buckwheat & Toasted Almond Morning Bowl
- Lunch: Garden Fresh Veggie & Egg White Frittata
- Snack: Roasted Pumpkin Seeds
- Dinner: Cauliflower Steak with Chimichurri Sauce
  - Side: Garlic Cauliflower "Rice"

Day 87:
- Breakfast: Mango & Coconut Tropical Escape Smoothie
- Lunch: Sesame Tofu & Crunchy Veggie Bowl
- Snack: Sweet Cinnamon Almond Mix
- Dinner: Vegan Lentil Meatballs in Tomato Basil Sauce
  - Side: Carrot and Zucchini Ribbons with Lemon Vinaigrette

Day 88:
- Breakfast: Summer Berries & Amaranth Cereal Bowl
- Lunch: Mediterranean Hummus & Tofu Salad
- Snack: Baked Kale Chips

- Dinner: Seared Tofu Steaks with Avocado-Wasabi Sauce
  - Side: Quinoa Salad with Cucumber and Avocado

Day 89:
- Breakfast: Warm Chia & Golden Hemp Heart Porridge
- Lunch: Kale & Roasted Chickpea Delight
- Snack: Zucchini Chips
- Dinner: Baked Tofu with Mango Salsa
  - Side: Mashed Sweet Potatoes with Cinnamon

Day 90:
- Breakfast: Crunchy Seed & Nut Morning Muesli
- Lunch: Herbed Egg & Veggie Fiesta
- Snack: Fresh Berry Salad
- Dinner: Vegetable Stir-Fry with Black Beans, Broccoli and Bell Peppers
  - Side: Roasted Brussels Sprouts with Balsamic Reduction

Day 91:
- Breakfast: Energizing Espresso & Oat Smoothie
- Lunch: Quinoa Power Breakfast Bowl
- Snack: Apple Cinnamon Chips
- Dinner: Grilled Portobello Mushrooms with Lentil Mint Pesto
  - Side: Sautéed Spinach with Pine Nuts and Raisins

WEEK 14

Day 92:
- Breakfast: Heart-Healthy Avocado & Mixed Berry Smoothie
- Lunch: Jackfruit Avocado Quinoa Salad
- Snack: Cucumber Cups with Herbed Almond Yogurt
- Dinner: Herb-Roasted Cauliflower Steaks
  - Side: Garlic Cauliflower "Rice"

Day 93:
- Breakfast: Velvety Blueberry Spinach Bliss Smoothie
- Lunch: Kale & Roasted Chickpea Delight
- Snack: Spiced Chickpea Nuts
- Dinner: Seared Tofu Steaks with Avocado-Wasabi Sauce
  - Side: Steamed Asparagus with Toasted Almonds

Day 94:
- Breakfast: Warm Chia & Golden Hemp Heart Porridge
- Lunch: Beetroot & Walnut Harmony
- Snack: Baked Kale Chips
- Dinner: Vegan Lentil Meatballs in Tomato Basil Sauce
  - Side: Quinoa Salad with Cucumber and Avocado

Day 95:
- Breakfast: Mango & Coconut Tropical Escape Smoothie
- Lunch: Sesame Tofu & Crunchy Veggie Bowl
- Snack: Nutty Stuffed Pears
- Dinner: Cauliflower Steak with Chimichurri Sauce
  - Side: Carrot and Zucchini Ribbons with Lemon Vinaigrette

Day 96:
- Breakfast: Apple Cinnamon Delight Shake
- Lunch: Mediterranean Hummus & Tofu Salad
- Snack: Roasted Pumpkin Seeds
- Dinner: Baked Tofu with Mango Salsa
  - Side: Mashed Sweet Potatoes with Cinnamon

Day 97:
- Breakfast: Energizing Espresso & Oat Smoothie
- Lunch: Herbed Egg & Veggie Fiesta
- Snack: Zucchini Chips
- Dinner: Vegetable Stir-Fry with Black Beans, Broccoli and Bell Peppers
  - Side: Roasted Brussels Sprouts with Balsamic Reduction

Day 98:
- Breakfast: Energizing Espresso & Oat Smoothie
- Lunch: Garden Fresh Veggie & Egg White Frittata
- Snack: Fresh Berry Salad
- Dinner: Grilled Portobello Mushrooms with Lentil Mint Pesto
  - Side: Sautéed Spinach with Pine Nuts and Raisins

WEEK 15

Day 99:
- Breakfast: Cinnamon Spiced Quinoa Breakfast Cereal
- Lunch: Broccoli Almond Bliss Salad
- Snack: Avocado Hummus
- Dinner: Vegan Shepherd's Pie with Lentils
  - Side: Steamed Asparagus with Toasted Almonds

Day 100:
- Breakfast: Buckwheat & Toasted Almond Morning Bowl
- Lunch: Garden Fresh Veggie & Egg White Frittata
- Snack: Roasted Pumpkin Seeds
- Dinner: Cauliflower Steak with Chimichurri Sauce
  - Side: Garlic Cauliflower "Rice"

Day 101:
- Breakfast: Mango & Coconut Tropical Escape Smoothie

- Lunch: Sesame Tofu & Crunchy Veggie Bowl
- Snack: Sweet Cinnamon Almond Mix
- Dinner: Vegan Lentil Meatballs in Tomato Basil Sauce
    - Side: Carrot and Zucchini Ribbons with Lemon Vinaigrette

Day 102:
- Breakfast: Summer Berries & Amaranth Cereal Bowl
- Lunch: Mediterranean Hummus & Tofu Salad
- Snack: Baked Kale Chips
- Dinner: Seared Tofu Steaks with Avocado-Wasabi Sauce
    - Side: Quinoa Salad with Cucumber and Avocado

Day 103:
- Breakfast: Warm Chia & Golden Hemp Heart Porridge
- Lunch: Kale & Roasted Chickpea Delight
- Snack: Zucchini Chips
- Dinner: Baked Tofu with Mango Salsa
    - Side: Mashed Sweet Potatoes with Cinnamon

Day 104:
- Breakfast: Crunchy Seed & Nut Morning Muesli
- Lunch: Herbed Egg & Veggie Fiesta
- Snack: Fresh Berry Salad
- Dinner: Vegetable Stir-Fry with Black Beans, Broccoli and Bell Peppers
    - Side: Roasted Brussels Sprouts with Balsamic Reduction

Day 105:
- Breakfast: Energizing Espresso & Oat Smoothie
- Lunch: Quinoa Power Breakfast Bowl
- Snack: Apple Cinnamon Chips
- Dinner: Grilled Portobello Mushrooms with Lentil Mint Pesto
    - Side: Sautéed Spinach with Pine Nuts and Raisins

WEEK 16

Day 106:
- Breakfast: Heart-Healthy Avocado & Mixed Berry Smoothie
- Lunch: Jackfruit Avocado Quinoa Salad
- Snack: Cucumber Cups with Herbed Almond Yogurt
- Dinner: Herb-Roasted Cauliflower Steaks
    - Side: Garlic Cauliflower "Rice"

Day 107:
- Breakfast: Velvety Blueberry Spinach Bliss Smoothie
- Lunch: Kale & Roasted Chickpea Delight
- Snack: Spiced Chickpea Nuts

- Dinner: Seared Tofu Steaks with Avocado-Wasabi Sauce
  - Side: Steamed Asparagus with Toasted Almonds

Day 108:
- Breakfast: Warm Chia & Golden Hemp Heart Porridge
- Lunch: Beetroot & Walnut Harmony
- Snack: Baked Kale Chips
- Dinner: Vegan Lentil Meatballs in Tomato Basil Sauce
  - Side: Quinoa Salad with Cucumber and Avocado

Day 109:
- Breakfast: Mango & Coconut Tropical Escape Smoothie
- Lunch: Sesame Tofu & Crunchy Veggie Bowl
- Snack: Nutty Stuffed Pears
- Dinner: Cauliflower Steak with Chimichurri Sauce
  - Side: Carrot and Zucchini Ribbons with Lemon Vinaigrette

Day 110:
- Breakfast: Apple Cinnamon Delight Shake
- Lunch: Mediterranean Hummus & Tofu Salad
- Snack: Roasted Pumpkin Seeds
- Dinner: Baked Tofu with Mango Salsa
  - Side: Mashed Sweet Potatoes with Cinnamon

Day 111:
- Breakfast: Energizing Espresso & Oat Smoothie
- Lunch: Herbed Egg & Veggie Fiesta
- Snack: Zucchini Chips
- Dinner: Vegetable Stir-Fry with Black Beans, Broccoli and Bell Peppers
  - Side: Roasted Brussels Sprouts with Balsamic Reduction

Day 112:
- Breakfast: Energizing Espresso & Oat Smoothie
- Lunch: Garden Fresh Veggie & Egg White Frittata
- Snack: Fresh Berry Salad
- Dinner: Grilled Portobello Mushrooms with Lentil Mint Pesto
  - Side: Sautéed Spinach with Pine Nuts and Raisins

WEEK 17

Day 113:
- Breakfast: Cinnamon Spiced Quinoa Breakfast Cereal
- Lunch: Broccoli Almond Bliss Salad
- Snack: Avocado Hummus
- Dinner: Vegan Shepherd's Pie with Lentils
  - Side: Steamed Asparagus with Toasted Almonds

Day 114:
- Breakfast: Buckwheat & Toasted Almond Morning Bowl
- Lunch: Garden Fresh Veggie & Egg White Frittata
- Snack: Roasted Pumpkin Seeds
- Dinner: Cauliflower Steak with Chimichurri Sauce
  - Side: Garlic Cauliflower "Rice"

Day 115:
- Breakfast: Mango & Coconut Tropical Escape Smoothie
- Lunch: Sesame Tofu & Crunchy Veggie Bowl
- Snack: Sweet Cinnamon Almond Mix
- Dinner: Vegan Lentil Meatballs in Tomato Basil Sauce
  - Side: Carrot and Zucchini Ribbons with Lemon Vinaigrette

Day 116:
- Breakfast: Summer Berries & Amaranth Cereal Bowl
- Lunch: Mediterranean Hummus & Tofu Salad
- Snack: Baked Kale Chips
- Dinner: Seared Tofu Steaks with Avocado-Wasabi Sauce
  - Side: Quinoa Salad with Cucumber and Avocado

Day 117:
- Breakfast: Warm Chia & Golden Hemp Heart Porridge
- Lunch: Kale & Roasted Chickpea Delight
- Snack: Zucchini Chips
- Dinner: Baked Tofu with Mango Salsa
  - Side: Mashed Sweet Potatoes with Cinnamon

Day 118:
- Breakfast: Crunchy Seed & Nut Morning Muesli
- Lunch: Herbed Egg & Veggie Fiesta
- Snack: Fresh Berry Salad
- Dinner: Vegetable Stir-Fry with Black Beans, Broccoli and Bell Peppers
  - Side: Roasted Brussels Sprouts with Balsamic Reduction

Day 119:
- Breakfast: Energizing Espresso & Oat Smoothie
- Lunch: Quinoa Power Breakfast Bowl
- Snack: Apple Cinnamon Chips
- Dinner: Grilled Portobello Mushrooms with Lentil Mint Pesto
  - Side: Sautéed Spinach with Pine Nuts and Raisins

WEEK 18

Day 120:
- Breakfast: Heart-Healthy Avocado & Mixed Berry Smoothie

- Lunch: Jackfruit Avocado Quinoa Salad
- Snack: Cucumber Cups with Herbed Almond Yogurt
- Dinner: Herb-Roasted Cauliflower Steaks
  - Side: Garlic Cauliflower "Rice"

Day 121:
- Breakfast: Velvety Blueberry Spinach Bliss Smoothie
- Lunch: Kale & Roasted Chickpea Delight
- Snack: Spiced Chickpea Nuts
- Dinner: Seared Tofu Steaks with Avocado-Wasabi Sauce
  - Side: Steamed Asparagus with Toasted Almonds

Day 122:
- Breakfast: Warm Chia & Golden Hemp Heart Porridge
- Lunch: Beetroot & Walnut Harmony
- Snack: Baked Kale Chips
- Dinner: Vegan Lentil Meatballs in Tomato Basil Sauce
  - Side: Quinoa Salad with Cucumber and Avocado

Day 123:
- Breakfast: Mango & Coconut Tropical Escape Smoothie
- Lunch: Sesame Tofu & Crunchy Veggie Bowl
- Snack: Nutty Stuffed Pears
- Dinner: Cauliflower Steak with Chimichurri Sauce
  - Side: Carrot and Zucchini Ribbons with Lemon Vinaigrette

Day 124:
- Breakfast: Apple Cinnamon Delight Shake
- Lunch: Mediterranean Hummus & Tofu Salad
- Snack: Roasted Pumpkin Seeds
- Dinner: Baked Tofu with Mango Salsa
  - Side: Mashed Sweet Potatoes with Cinnamon

Day 125:
- Breakfast: Energizing Espresso & Oat Smoothie
- Lunch: Herbed Egg & Veggie Fiesta
- Snack: Zucchini Chips
- Dinner: Vegetable Stir-Fry with Black Beans, Broccoli and Bell Peppers
  - Side: Roasted Brussels Sprouts with Balsamic Reduction

Day 126:
- Breakfast: Energizing Espresso & Oat Smoothie
- Lunch: Garden Fresh Veggie & Egg White Frittata
- Snack: Fresh Berry Salad
- Dinner: Grilled Portobello Mushrooms with Lentil Mint Pesto
  - Side: Sautéed Spinach with Pine Nuts and Raisins

WEEK 19

Day 127:
- Breakfast: Cinnamon Spiced Quinoa Breakfast Cereal
- Lunch: Broccoli Almond Bliss Salad
- Snack: Avocado Hummus
- Dinner: Vegan Shepherd's Pie with Lentils
  - Side: Steamed Asparagus with Toasted Almonds

Day 128:
- Breakfast: Buckwheat & Toasted Almond Morning Bowl
- Lunch: Garden Fresh Veggie & Egg White Frittata
- Snack: Roasted Pumpkin Seeds
- Dinner: Cauliflower Steak with Chimichurri Sauce
  - Side: Garlic Cauliflower "Rice"

Day 129:
- Breakfast: Mango & Coconut Tropical Escape Smoothie
- Lunch: Sesame Tofu & Crunchy Veggie Bowl
- Snack: Sweet Cinnamon Almond Mix
- Dinner: Vegan Lentil Meatballs in Tomato Basil Sauce
  - Side: Carrot and Zucchini Ribbons with Lemon Vinaigrette

Day 130:
- Breakfast: Summer Berries & Amaranth Cereal Bowl
- Lunch: Mediterranean Hummus & Tofu Salad
- Snack: Baked Kale Chips
- Dinner: Seared Tofu Steaks with Avocado-Wasabi Sauce
  - Side: Quinoa Salad with Cucumber and Avocado

Day 131:
- Breakfast: Warm Chia & Golden Hemp Heart Porridge
- Lunch: Kale & Roasted Chickpea Delight
- Snack: Zucchini Chips
- Dinner: Baked Tofu with Mango Salsa
  - Side: Mashed Sweet Potatoes with Cinnamon

Day 132:
- Breakfast: Crunchy Seed & Nut Morning Muesli
- Lunch: Herbed Egg & Veggie Fiesta
- Snack: Fresh Berry Salad
- Dinner: Vegetable Stir-Fry with Black Beans, Broccoli and Bell Peppers
  - Side: Roasted Brussels Sprouts with Balsamic Reduction

Day 133:
- Breakfast: Energizing Espresso & Oat Smoothie

- Lunch: Quinoa Power Breakfast Bowl
- Snack: Apple Cinnamon Chips
- Dinner: Grilled Portobello Mushrooms with Lentil Mint Pesto
  - Side: Sautéed Spinach with Pine Nuts and Raisins

WEEK 20

Day 134:
- Breakfast: Heart-Healthy Avocado & Mixed Berry Smoothie
- Lunch: Jackfruit Avocado Quinoa Salad
- Snack: Cucumber Cups with Herbed Almond Yogurt
- Dinner: Herb-Roasted Cauliflower Steaks
  - Side: Garlic Cauliflower "Rice"

Day 135:
- Breakfast: Velvety Blueberry Spinach Bliss Smoothie
- Lunch: Kale & Roasted Chickpea Delight
- Snack: Spiced Chickpea Nuts
- Dinner: Seared Tofu Steaks with Avocado-Wasabi Sauce
  - Side: Steamed Asparagus with Toasted Almonds

Day 136:
- Breakfast: Warm Chia & Golden Hemp Heart Porridge
- Lunch: Beetroot & Walnut Harmony
- Snack: Baked Kale Chips
- Dinner: Vegan Lentil Meatballs in Tomato Basil Sauce
  - Side: Quinoa Salad with Cucumber and Avocado

Day 137:
- Breakfast: Mango & Coconut Tropical Escape Smoothie
- Lunch: Sesame Tofu & Crunchy Veggie Bowl
- Snack: Nutty Stuffed Pears
- Dinner: Cauliflower Steak with Chimichurri Sauce
  - Side: Carrot and Zucchini Ribbons with Lemon Vinaigrette

Day 138:
- Breakfast: Apple Cinnamon Delight Shake
- Lunch: Mediterranean Hummus & Tofu Salad
- Snack: Roasted Pumpkin Seeds
- Dinner: Baked Tofu with Mango Salsa
  - Side: Mashed Sweet Potatoes with Cinnamon

Day 139:
- Breakfast: Energizing Espresso & Oat Smoothie
- Lunch: Herbed Egg & Veggie Fiesta
- Snack: Zucchini Chips

- Dinner: Vegetable Stir-Fry with Black Beans, Broccoli and Bell Peppers
  - Side: Roasted Brussels Sprouts with Balsamic Reduction

Day 140:
- Breakfast: Energizing Espresso & Oat Smoothie
- Lunch: Garden Fresh Veggie & Egg White Frittata
- Snack: Fresh Berry Salad
- Dinner: Grilled Portobello Mushrooms with Lentil Mint Pesto
  - Side: Sautéed Spinach with Pine Nuts and Raisins

WEEK 21

Day 141:
- Breakfast: Cinnamon Spiced Quinoa Breakfast Cereal
- Lunch: Broccoli Almond Bliss Salad
- Snack: Avocado Hummus
- Dinner: Vegan Shepherd's Pie with Lentils
  - Side: Steamed Asparagus with Toasted Almonds

Day 142:
- Breakfast: Buckwheat & Toasted Almond Morning Bowl
- Lunch: Garden Fresh Veggie & Egg White Frittata
- Snack: Roasted Pumpkin Seeds
- Dinner: Cauliflower Steak with Chimichurri Sauce
  - Side: Garlic Cauliflower "Rice"

Day 143:
- Breakfast: Mango & Coconut Tropical Escape Smoothie
- Lunch: Sesame Tofu & Crunchy Veggie Bowl
- Snack: Sweet Cinnamon Almond Mix
- Dinner: Vegan Lentil Meatballs in Tomato Basil Sauce
  - Side: Carrot and Zucchini Ribbons with Lemon Vinaigrette

Day 144:
- Breakfast: Summer Berries & Amaranth Cereal Bowl
- Lunch: Mediterranean Hummus & Tofu Salad
- Snack: Baked Kale Chips
- Dinner: Seared Tofu Steaks with Avocado-Wasabi Sauce
  - Side: Quinoa Salad with Cucumber and Avocado

Day 145:
- Breakfast: Warm Chia & Golden Hemp Heart Porridge
- Lunch: Kale & Roasted Chickpea Delight
- Snack: Zucchini Chips
- Dinner: Baked Tofu with Mango Salsa
  - Side: Mashed Sweet Potatoes with Cinnamon

Day 146:
- Breakfast: Crunchy Seed & Nut Morning Muesli
- Lunch: Herbed Egg & Veggie Fiesta
- Snack: Fresh Berry Salad
- Dinner: Vegetable Stir-Fry with Black Beans, Broccoli and Bell Peppers
  - Side: Roasted Brussels Sprouts with Balsamic Reduction

Day 147:
- Breakfast: Energizing Espresso & Oat Smoothie
- Lunch: Quinoa Power Breakfast Bowl
- Snack: Apple Cinnamon Chips
- Dinner: Grilled Portobello Mushrooms with Lentil Mint Pesto
  - Side: Sautéed Spinach with Pine Nuts and Raisins

WEEK 22

Day 148:
- Breakfast: Heart-Healthy Avocado & Mixed Berry Smoothie
- Lunch: Jackfruit Avocado Quinoa Salad
- Snack: Cucumber Cups with Herbed Almond Yogurt
- Dinner: Herb-Roasted Cauliflower Steaks
  - Side: Garlic Cauliflower "Rice"

Day 149:
- Breakfast: Velvety Blueberry Spinach Bliss Smoothie
- Lunch: Kale & Roasted Chickpea Delight
- Snack: Spiced Chickpea Nuts
- Dinner: Seared Tofu Steaks with Avocado-Wasabi Sauce
  - Side: Steamed Asparagus with Toasted Almonds

Day 150:
- Breakfast: Warm Chia & Golden Hemp Heart Porridge
- Lunch: Beetroot & Walnut Harmony
- Snack: Baked Kale Chips
- Dinner: Vegan Lentil Meatballs in Tomato Basil Sauce
  - Side: Quinoa Salad with Cucumber and Avocado

Day 151:
- Breakfast: Mango & Coconut Tropical Escape Smoothie
- Lunch: Sesame Tofu & Crunchy Veggie Bowl
- Snack: Nutty Stuffed Pears
- Dinner: Cauliflower Steak with Chimichurri Sauce
  - Side: Carrot and Zucchini Ribbons with Lemon Vinaigrette

Day 152:
- Breakfast: Apple Cinnamon Delight Shake

- Lunch: Mediterranean Hummus & Tofu Salad
- Snack: Roasted Pumpkin Seeds
- Dinner: Baked Tofu with Mango Salsa
  - Side: Mashed Sweet Potatoes with Cinnamon

Day 153:
- Breakfast: Energizing Espresso & Oat Smoothie
- Lunch: Herbed Egg & Veggie Fiesta
- Snack: Zucchini Chips
- Dinner: Vegetable Stir-Fry with Black Beans, Broccoli and Bell Peppers
  - Side: Roasted Brussels Sprouts with Balsamic Reduction

Day 154:
- Breakfast: Energizing Espresso & Oat Smoothie
- Lunch: Garden Fresh Veggie & Egg White Frittata
- Snack: Fresh Berry Salad
- Dinner: Grilled Portobello Mushrooms with Lentil Mint Pesto
  - Side: Sautéed Spinach with Pine Nuts and Raisins

WEEK 23

Day 155:
- **Breakfast:** Velvety Blueberry Spinach Bliss Smoothie
- **Lunch:** Mediterranean Hummus & Tofu Salad
- **Snack:** Zucchini Chips
- **Dinner:** Grilled Portobello Mushrooms with Lentil Mint Pesto
  - **Side:** Quinoa Salad with Cucumber and Avocado

Day 156:
- **Breakfast:** Cinnamon Spiced Quinoa Breakfast Cereal
- **Lunch:** Vegan Jambalaya with Jackfruit
- **Snack:** Fresh Berry Salad
- **Dinner:** Vegan Lentil Meatballs in Tomato Basil Sauce
  - **Side:** Steamed Asparagus with Toasted Almonds

Day 157:
- **Breakfast:** Mango & Coconut Tropical Escape Smoothie
- **Lunch:** Garden Fresh Veggie & Egg White Frittata
- **Snack:** Apple Cinnamon Chips
- **Dinner:** Cauliflower Steak with Chimichurri Sauce
  - **Side:** Mashed Sweet Potatoes with Cinnamon

Day 158:
- **Breakfast:** Heart-Healthy Avocado & Mixed Berry Smoothie
- **Lunch:** Jackfruit Avocado Quinoa Salad
- **Snack:** Spiced Chickpea Nuts

- **Dinner:** Seared Tofu Steaks with Avocado-Wasabi Sauce
  - **Side:** Carrot and Zucchini Ribbons with Lemon Vinaigrette

Day 159:
- **Breakfast:** Buckwheat & Toasted Almond Morning Bowl
- **Lunch:** Broccoli Almond Bliss Salad
- **Snack:** Baked Kale Chips
- **Dinner:** Mushroom Stroganoff with Cashew Cream
  - **Side:** Roasted Brussels Sprouts with Balsamic Reduction

Day 160:
- **Breakfast:** Protein-Packed Peanut Butter & Strawberry Smoothie
- **Lunch:** Lentil & Avocado Power Bowl
- **Snack:** Roasted Pumpkin Seeds
- **Dinner:** Baked Tofu with Mango Salsa
  - **Side:** Garlic Cauliflower "Rice"

Day 161:
- **Breakfast:** Energizing Espresso & Oat Smoothie
- **Lunch:** Quinoa and Black Bean Stuffed Bell Peppers
- **Snack:** Nutty Stuffed Pears
- **Dinner:** Vegan Shepherd's Pie with Lentils
  - **Side:** Sautéed Spinach with Pine Nuts and Raisins

WEEK 24

Day 162:
- **Breakfast:** Crunchy Seed & Nut Morning Muesli
- **Lunch:** Savory Tofu & Spinach Scramble
- **Snack:** Cucumber Cups with Herbed Yogurt
- **Dinner:** Herb-Roasted Cauliflower Breast
  - **Side:** Steamed Asparagus with Toasted Almonds

Day 163:
- **Breakfast:** Apple Cinnamon Delight Shake
- **Lunch:** Sesame Tofu & Crunchy Veggie Bowl
- **Snack:** Apple Cinnamon Chips
- **Dinner:** Zucchini Noodle Pad Thai
  - **Side:** Garlic Cauliflower "Rice"

Day 164:
- **Breakfast:** Mango & Coconut Tropical Escape Smoothie
- **Lunch:** Quinoa Power Breakfast Bowl with Soft Boiled Egg
- **Snack:** Spiced Chickpea Nuts
- **Dinner:** Vegan Lentil Salad Sandwich on Whole Grain Bread
  - **Side:** Carrot and Zucchini Ribbons with Lemon Vinaigrette

Day 165:
- **Breakfast:** Velvety Blueberry Spinach Bliss Smoothie
- **Lunch:** Jackfruit Avocado Quinoa Salad
- **Snack:** Baked Kale Chips
- **Dinner:** Vegetable Stir-Fry with Black Beans, Broccoli and Bell Peppers
  - **Side:** Mashed Sweet Potatoes with Cinnamon

Day 166:
- **Breakfast:** Energizing Espresso & Oat Smoothie
- **Lunch:** Mediterranean Hummus & Tofu Salad
- **Snack:** Roasted Pumpkin Seeds
- **Dinner:** Grilled Portobello Mushrooms with Lentil Mint Pesto
  - **Side:** Roasted Brussels Sprouts with Balsamic Reduction

Day 167:
- **Breakfast:** Heart-Healthy Avocado & Mixed Berry Smoothie
- **Lunch:** Lentil & Avocado Power Bowl
- **Snack:** Zucchini Chips
- **Dinner:** Cauliflower Steak with Chimichurri Sauce
  - **Side:** Quinoa Salad with Cucumber and Avocado

Day 168:
- **Breakfast:** Cinnamon Spiced Quinoa Breakfast Cereal
- **Lunch:** Vegan Jambalaya with Jackfruit
- **Snack:** Fresh Berry Salad
- **Dinner:** Baked Tofu with Mango Salsa
  - **Side:** Steamed Asparagus with Toasted Almonds

Day 169:
- **Breakfast:** Buckwheat & Toasted Almond Morning Bowl
- **Lunch:** Garden Fresh Veggie & Egg White Frittata
- **Snack:** Spiced Pear Slices
- **Dinner:** Vegan Shepherd's Pie with Lentils
  - **Side:** Sautéed Spinach with Pine Nuts and Raisins

WEEK 25

Day 170:
- **Breakfast:** Mango & Coconut Tropical Escape Smoothie
- **Lunch:** Broccoli Almond Bliss Salad
- **Snack:** Nutty Stuffed Pears
- **Dinner:** Mushroom Stroganoff with Cashew Cream
  - **Side:** Garlic Cauliflower "Rice"

Day 171:
- **Breakfast:** Crunchy Seed & Nut Morning Muesli

- **Lunch:** Sesame Tofu & Crunchy Veggie Bowl
- **Snack:** Cucumber Cups with Herbed Yogurt
- **Dinner:** Vegetable Stir-Fry with Black Beans, Broccoli and Bell Peppers
  - **Side:** Carrot and Zucchini Ribbons with Lemon Vinaigrette

Day 172:
- **Breakfast:** Warm Chia & Golden Hemp Heart Porridge
- **Lunch:** Quinoa and Black Bean Stuffed Bell Peppers
- **Snack:** Apple Cinnamon Chips
- **Dinner:** Seared Tofu Steaks with Avocado-Wasabi Sauce
  - **Side:** Quinoa Salad with Cucumber and Avocado

Day 173:
- **Breakfast:** Apple Cinnamon Delight Shake
- **Lunch:** Jackfruit Avocado Quinoa Salad
- **Snack:** Spiced Chickpea Nuts
- **Dinner:** Grilled Portobello Mushrooms with Lentil Mint Pesto
  - **Side:** Mashed Sweet Potatoes with Cinnamon

Day 174:
- **Breakfast:** Velvety Blueberry Spinach Bliss Smoothie
- **Lunch:** Mediterranean Hummus & Tofu Salad
- **Snack:** Baked Kale Chips
- **Dinner:** Cauliflower Steak with Chimichurri Sauce
  - **Side:** Roasted Brussels Sprouts with Balsamic Reduction

Day 175:
- **Breakfast:** Energizing Espresso & Oat Smoothie
- **Lunch:** Lentil & Avocado Power Bowl
- **Snack:** Roasted Pumpkin Seeds
- **Dinner:** Vegan Lentil Salad Sandwich on Whole Grain Bread
  - **Side:** Steamed Asparagus with Toasted Almonds

Day 176:
- **Breakfast:** Heart-Healthy Avocado & Mixed Berry Smoothie
- **Lunch:** Broccoli Almond Bliss Salad
- **Snack:** Zucchini Chips
- **Dinner:** Baked Tofu with Mango Salsa
  - **Side:** Garlic Cauliflower "Rice"

WEEK 26

Day 177:
- **Breakfast:** Crunchy Seed & Nut Morning Muesli
- **Lunch:** Vegan Jambalaya with Jackfruit
- **Snack:** Nutty Stuffed Pears

- **Dinner:** Vegetable Stir-Fry with Black Beans, Broccoli and Bell Peppers
  - **Side:** Quinoa Salad with Cucumber and Avocado

Day 178:
- **Breakfast:** Mango & Coconut Tropical Escape Smoothie
- **Lunch:** Garden Fresh Veggie & Egg White Frittata
- **Snack:** Fresh Berry Salad
- **Dinner:** Grilled Portobello Mushrooms with Lentil Mint Pesto
  - **Side:** Carrot and Zucchini Ribbons with Lemon Vinaigrette

Day 179:
- **Breakfast:** Warm Chia & Golden Hemp Heart Porridge
- **Lunch:** Sesame Tofu & Crunchy Veggie Bowl
- **Snack:** Apple Cinnamon Chips
- **Dinner:** Mushroom Stroganoff with Cashew Cream
  - **Side:** Mashed Sweet Potatoes with Cinnamon

Day 180:
- **Breakfast:** Apple Cinnamon Delight Shake
- **Lunch:** Mediterranean Hummus & Tofu Salad
- **Snack:** Roasted Pumpkin Seeds
- **Dinner:** Vegan Shepherd's Pie with Lentils
  - **Side:** Steamed Asparagus with Toasted Almonds

Day 181:
- **Breakfast:** Velvety Blueberry Spinach Bliss Smoothie
- **Lunch:** Quinoa and Black Bean Stuffed Bell Peppers
- **Snack:** Spiced Chickpea Nuts
- **Dinner:** Cauliflower Steak with Chimichurri Sauce
  - **Side:** Garlic Cauliflower "Rice"

Day 182:
- **Breakfast:** Energizing Espresso & Oat Smoothie
- **Lunch:** Broccoli Almond Bliss Salad
- **Snack:** Zucchini Chips
- **Dinner:** Baked Tofu with Mango Salsa
  - **Side:** Roasted Brussels Sprouts with Balsamic Reduction

Day 183:
- **Breakfast:** Heart-Healthy Avocado & Mixed Berry Smoothie
- **Lunch:** Lentil & Avocado Power Bowl
- **Snack:** Cucumber Cups with Herbed Yogurt
- **Dinner:** Seared Tofu Steaks with Avocado-Wasabi Sauce
  - **Side:** Carrot and Zucchini Ribbons with Lemon Vinaigrette

WEEK 27

Day 184:
- **Breakfast:** Cinnamon Spiced Quinoa Breakfast Cereal
- **Lunch:** Jackfruit Avocado Quinoa Salad
- **Snack:** Baked Kale Chips
- **Dinner:** Grilled Portobello Mushrooms with Lentil Mint Pesto
  - **Side:** Quinoa Salad with Cucumber and Avocado

Day 185:
- **Breakfast:** Buckwheat & Toasted Almond Morning Bowl
- **Lunch:** Garden Fresh Veggie & Egg White Frittata
- **Snack:** Spiced Pear Slices
- **Dinner:** Vegan Lentil Salad Sandwich on Whole Grain Bread
  - **Side:** Mashed Sweet Potatoes with Cinnamon

Day 186:
- **Breakfast:** Mango & Coconut Tropical Escape Smoothie
- **Lunch:** Sesame Tofu & Crunchy Veggie Bowl
- **Snack:** Nutty Stuffed Pears
- **Dinner:** Vegetable Stir-Fry with Black Beans, Broccoli and Bell Peppers
  - **Side:** Garlic Cauliflower "Rice"

Day 187:
- **Breakfast:** Velvety Blueberry Spinach Bliss Smoothie
- **Lunch:** Mediterranean Hummus & Tofu Salad
- **Snack:** Fresh Berry Salad
- **Dinner:** Cauliflower Steak with Chimichurri Sauce
  - **Side:** Roasted Brussels Sprouts with Balsamic Reduction

Day 188:
- **Breakfast:** Energizing Espresso & Oat Smoothie
- **Lunch:** Quinoa and Black Bean Stuffed Bell Peppers
- **Snack:** Apple Cinnamon Chips
- **Dinner:** Baked Tofu with Mango Salsa
  - **Side:** Steamed Asparagus with Toasted Almonds

Day 189:
- **Breakfast:** Crunchy Seed & Nut Morning Muesli
- **Lunch:** Vegan Jambalaya with Jackfruit
- **Snack:** Roasted Pumpkin Seeds
- **Dinner:** Mushroom Stroganoff with Cashew Cream
  - **Side:** Carrot and Zucchini Ribbons with Lemon Vinaigrette

Day 190:
- **Breakfast:** Warm Chia & Golden Hemp Heart Porridge

- **Lunch:** Broccoli Almond Bliss Salad
- **Snack:** Zucchini Chips
- **Dinner:** Grilled Portobello Mushrooms with Lentil Mint Pesto
  - **Side:** Quinoa Salad with Cucumber and Avocado

WEEK 28

Day 191:
- **Breakfast:** Cinnamon Spiced Quinoa Breakfast Cereal
- **Lunch:** Jackfruit Avocado Quinoa Salad
- **Snack:** Baked Kale Chips
- **Dinner:** Vegan Shepherd's Pie with Lentils
  - **Side:** Steamed Asparagus with Toasted Almonds

Day 192:
- **Breakfast:** Buckwheat & Toasted Almond Morning Bowl
- **Lunch:** Garden Fresh Veggie & Egg White Frittata
- **Snack:** Spiced Pear Slices
- **Dinner:** Cauliflower Steak with Chimichurri Sauce
  - **Side:** Mashed Sweet Potatoes with Cinnamon

Day 193:
- **Breakfast:** Mango & Coconut Tropical Escape Smoothie
- **Lunch:** Sesame Tofu & Crunchy Veggie Bowl
- **Snack:** Nutty Stuffed Pears
- **Dinner:** Vegetable Stir-Fry with Black Beans, Broccoli and Bell Peppers
  - **Side:** Garlic Cauliflower "Rice"

Day 194:
- **Breakfast:** Velvety Blueberry Spinach Bliss Smoothie
- **Lunch:** Mediterranean Hummus & Tofu Salad
- **Snack:** Fresh Berry Salad
- **Dinner:** Baked Tofu with Mango Salsa
  - **Side:** Roasted Brussels Sprouts with Balsamic Reduction

Day 195:
- **Breakfast:** Energizing Espresso & Oat Smoothie
- **Lunch:** Quinoa and Black Bean Stuffed Bell Peppers
- **Snack:** Apple Cinnamon Chips
- **Dinner:** Grilled Portobello Mushrooms with Lentil Mint Pesto
  - **Side:** Steamed Asparagus with Toasted Almonds

Day 196:
- **Breakfast:** Crunchy Seed & Nut Morning Muesli
- **Lunch:** Vegan Jambalaya with Jackfruit
- **Snack:** Roasted Pumpkin Seeds

- **Dinner:** Mushroom Stroganoff with Cashew Cream
  - **Side:** Carrot and Zucchini Ribbons with Lemon Vinaigrette

Day 197:
- **Breakfast:** Warm Chia & Golden Hemp Heart Porridge
- **Lunch:** Broccoli Almond Bliss Salad
- **Snack:** Zucchini Chips
- **Dinner:** Cauliflower Steak with Chimichurri Sauce
  - **Side:** Quinoa Salad with Cucumber and Avocado

CHAPTER 9
COOKING CONVERSIONS

Here are some common measurement conversions:

Volume Equivalents (Liquid)

US Standard	US Standard (oz.)	Metric (approximate)
2 tbsps.	1 fl. oz.	30 milliliter
¼ cup	2 fl. oz.	60 milliliter
½ cup	4 fl. oz.	120 milliliter
1 cup	8 fl. oz.	240 milliliter
1½ cups	12 fl. oz.	355 milliliter
2 cups or 1 pint	16 fl. oz.	475 milliliter
4 cups or 1 quart	32 fl. oz.	1 Liter
1 gallon	128 fl. oz.	4 Liter

Volume Equivalents (Dry)

US Standard	Metric (approximate)
⅛ tsp.	0.5 milliliter
¼ tsp.	1 milliliter
½ tsp.	2 milliliter
¾ tsp.	4 milliliter
1 tsp.	5 milliliter
1 tbsp.	15 milliliter
¼ cup	59 milliliter
⅓ cup	79 milliliter
½ cup	118 milliliter
⅔ cup	156 milliliter
¾ cup	177 milliliter
1 cup	235 milliliter

2 cups or 1 pint	475 milliliter
3 cups	700 milliliter
4 cups or 1 quart	1 Liter

Oven Temperatures

Fahrenheit (F)	Celsius (C) (approximate)
250 deg.F	120 deg.F
300 deg.F	150 deg.F
325 deg.F	165 deg.F
350 deg.F	180 deg.F
375 deg.F	190 deg.F
400 deg.F	200 deg.F
425 deg.F	220 deg.F
450 deg.F	230 deg.F

Weight Equivalents

US Standard	Metric (approximate)
1 tbsp.	15 g
½ oz.	15 g
1 oz.	30 g
2 oz.	60 g
4 oz.	115 g
8 oz.	225 g
12 oz.	340 g
16 oz. or 1 lb.	455 g

CONCLUSION

As we wrap up this culinary journey in the Dr. Barbara Diabetes Cookbook and Meal Plan, it feels only right to take a moment to reflect on the essence of what we've tried to create here. This isn't just a collection of recipes—it's a testament to the power of food as medicine, a belief deeply rooted in Dr. Barbara O'Neill's holistic approach to health and well-being.

Navigating the daily challenges of diabetes isn't merely about monitoring blood sugar levels or managing dietary dos and don'ts; it's about rediscovering joy in the simple act of nourishing your body. Each recipe in this book has been crafted with love and care, designed not only to meet the nutritional needs of those managing diabetes but also to delight the palate and satisfy the soul.

Imagine starting your day with a Velvety Blueberry Spinach Bliss Smoothie—a blend so rich in antioxidants and vibrant colors that it feels like you're drinking in the very essence of vitality. Or consider the comfort of sitting down to a dinner of Grilled Eggplant and Pesto Panini, where every bite is a reminder of the lushness of a well-tended garden.

Throughout this book, we've avoided processed foods not just for their health implications, but to steer closer to the natural world, a principle Dr. O'Neill champions fervently. The fresh, unprocessed ingredients like hearty greens, wholesome grains, and succulent fruits are a call back to the basics, to a time when food was about community and connection—values that Dr. O'Neill holds dear.

Each section of this cookbook has been thoughtfully curated to ensure that it serves not just as a meal plan but as a guide to sustainable, healthy living. We've included detailed nutritional information to empower you with the knowledge to make informed decisions about your diet. We've steered clear of the fleeting allure of quick fixes, focusing instead on creating a balanced, sustainable lifestyle that enhances your well-being without sacrificing flavor.

The journey through the pages of this cookbook is much like wandering through a lush garden; it's about pausing to savor the bright burst of a ripe berry, the earthy aroma of fresh herbs, and the comforting simplicity of grains. It's about transforming the act of eating from a mundane task into a daily celebration of life's flavorful abundance.

As you continue to explore these recipes and integrate them into your daily routine, remember that each meal is an opportunity to nurture not only your body but also your spirit. The path to managing diabetes is as much about what you eat as it is about cultivating a sense of peace and enjoyment with every meal.

Let this cookbook be a companion on your journey to health, a source of inspiration when meals seem monotonous, and a beacon of hope that even within the constraints of dietary needs, there is a vast landscape of flavors to explore. Embrace this adventure with an open heart and a willing spirit, and let the art of cooking bring joy back into your kitchen.

In conclusion, may these recipes bring you not only health and balance but also a deeper appreciation for the nourishing power of nature's bounty, inspired by the wisdom of Dr. Barbara O'Neill. Here's to eating well, living well, and thriving with every chosen ingredient and prepared meal.

Dear Reader,

Our team has dedicated considerable time and effort to develop this book, with the goal of providing a high-quality and insightful publication. Your review on Amazon would be greatly valued and crucial in helping us reach more readers. We deeply appreciate your support and are sincerely thankful for any feedback you decide to offer!

ANALYTIC INDEX

Almond Butter & Mixed Berry Smoothie, 11
Almond Butter Energy Balls, 73
Antioxidant-Rich Green Tea & Kiwi Smoothie, 9
Apple Cinnamon Chips, 70
Apple Cinnamon Delight Shake, 10
Avocado & Quinoa Salad Wrap, 34
Avocado and Tomato Cucumber Cups, 75
Avocado Hummus, 66
Avocado Vanilla Mousse, 72
Baked Kale Chips, 67
Baked Tofu with Mango Salsa, 58
Beetroot & Walnut Harmony, 26
Berry Bliss Almond Yogurt Overnight Oats, 21
Broccoli Almond Bliss Salad, 27
Buckwheat & Toasted Almond Morning Bowl, 13
Carrot and Zucchini Ribbons with Lemon Vinaigrette, 62
Carrot Cake Balls, 72
Cauliflower Steak with Chimichurri Sauce, 50
Chia and Berry Parfait, 70
Chickpea Caesar Wrap, 35
Chickpea Salad Sandwich, 39
Cinnamon Apple Pie Overnight Oats, 21
Cinnamon Flaxseed Pudding, 69
Cinnamon Nut Snack Mix, 74
Cinnamon Spiced Quinoa Breakfast Cereal, 13
Creamy Almond Butter & Banana Power Shake, 8
Creamy Carrot and Ginger Soup, 41
Crunchy Seed & Nut Morning Muesli, 12
Cucumber Cups with Herbed Almond Yogurt, 66
Energizing Espresso & Oat Smoothie, 11
Flaxseed and Blueberry Parfait, 74
Fresh Berry Salad, 69
Garden Fresh Veggie & Egg White Frittata, 17
Garlic Cauliflower "Rice", 62
Grilled Corn on the Cob with Chili Lime Dressing, 63
Grilled Eggplant and Fresh Pesto Panini, 37
Grilled Portobello Mushrooms with Lentil Mint Pesto, 59
Grilled Tofu with Lemon-Dill Sauce, 54
Heart-Healthy Avocado & Mixed Berry Smoothie, 9
Herb and Garlic Mushroom Caps, 68
Herbed Egg & Veggie Fiesta, 32
Herb-Roasted Cauliflower Steaks, 54
Homemade Vegetable Broth, 40
Homemade Veggie Burger Sandwich, 39
Homemade Whole Grain Tortilla Recipe, 33
Jackfruit Avocado Quinoa Salad, 23
Kale & Roasted Chickpea Delight, 24
Lentil & Avocado Power Bowl, 28
Lentil and Spinach Soup, 43
Luxurious Avocado & Chickpea Toast, 17
Mango & Coconut Tropical Escape Smoothie, 10
Mashed Sweet Potatoes with Cinnamon, 61
Mediterranean Hummus & Tofu Salad, 29
Mushroom and Barley Soup, 43
Mushroom Stroganoff with Cashew Cream, 51
Nutty Almond Yogurt & Fresh Fruit Parfait, 18
Nutty Stuffed Pears, 71
Pecan & Barley Harvest Bowl, 15
Pumpkin Seed and Sunflower Seed Mix, 75
Quinoa and Black Bean Stuffed Bell Peppers, 57
Quinoa Power Breakfast Bowl with Soft Boiled Egg, 20
Quinoa Salad with Cucumber and Avocado, 60
Ratatouille with Herbed Polenta, 52

Roasted Brussels Sprouts with Balsamic Reduction, 61
Roasted Pepper and Hummus Wrap, 36
Roasted Pumpkin Seeds, 67
Roasted Tomato Basil Soup, 40
Sautéed Spinach with Pine Nuts and Raisins, 63
Savory Chickpea and Vegetable Soup, 42
Savory Roasted Chickpeas, 73
Savory Tofu & Spinach Scramble, 16
Seared Tofu Steaks with Avocado-Wasabi Sauce, 55
Sesame Tofu & Crunchy Veggie Bowl, 25
Simple Jackfruit and Wild Rice Soup, 45
Southwest Tofu & Black Bean Lettuce Wrap, 19
Spaghetti Squash with Tomato Basil Sauce, 52
Spiced Chickpea Nuts, 65
Spiced Pear Slices, 76
Steamed Asparagus with Toasted Almonds, 60
Stuffed Bell Peppers with Quinoa and Black Beans, 47
Stuffed Bell Peppers with Quinoa and Vegetables, 68
Summer Berries & Amaranth Cereal Bowl, 14
Sweet Cinnamon Almond Mix, 71
Sweet Potato & Black Bean Breakfast Hash, 18
Sweet Potato and Almond Butter Slices (Baked), 76
Sweet Potato and Coconut Soup, 44
Tropical Coconut & Spelt Cereal, 15
Tropical Pineapple & Almond Yogurt Bowl, 16
Tropical Tofu & Fruit Salad, 31
Vegan & Savory Black Bean Wrap, 30
Vegan Corn and Potato Chowder, 46
Vegan Jambalaya with Jackfruit, 51
Vegan Lentil Meatballs in Tomato Basil Sauce, 56
Vegan Lentil Salad Sandwich on Whole Grain Bread, 38
Vegan Shepherd's Pie with Lentils, 48
Vegetable Stir-Fry with Black Beans, Broccoli and Bell Peppers, 59
Velvety Blueberry Spinach Bliss Smoothie, 8
Walnut & Oat Bran Heart-Smart Cereal, 14
Warm Chia & Golden Hemp Heart Porridge, 12
Zesty Black Bean and Corn Wrap, 34
Zucchini Chips, 65
Zucchini Noodle Pad Thai, 49

Made in United States
North Haven, CT
20 September 2024